URBAN LIVING LABS

C000263718

All cities face a pressing challenge – how can they provide economic prosperity and social cohesion while achieving environmental sustainability? In response, new collaborations are emerging in the form of urban living labs – sites devised to design, test and learn from social and technical innovation in real time. The aim of this volume is to examine, inform and advance the governance of sustainability transitions through urban living labs. Notably, urban living labs are proliferating rapidly across the globe as a means through which public and private actors are testing innovations in buildings, transport and energy systems. Yet despite the experimentation taking place on the ground, we lack systematic learning and international comparison across urban and national contexts about their impacts and effectiveness. We have limited knowledge on how good practice can be scaled up to achieve the transformative change required. This book brings together leading international researchers within a systematic comparative framework for evaluating the design, practices and processes of urban living labs to enable the comparative analysis of their potential and limits. It provides new insights into the governance of urban sustainability and how to improve the design and implementation of urban living labs in order to realise their potential.

Simon Marvin is Director of the Urban Institute and Professor in the Department of Geography at the University of Sheffield, UK.

Harriet Bulkeley is Professor of Geography at Durham University, UK.

Lindsay Mai is a Postdoctoral Research Associate in the Department of Geography at Durham University, UK.

Kes McCormick is an Associate Professor and Assistant Head at the International Institute for Industrial Environmental Economics (IIIEE) at Lund University, Sweden.

Yuliya Voytenko Palgan is an Assistant Professor at Lund University, Sweden.

URBAN LIVING LABS

Experimenting with City Futures

Edited by
Simon Marvin, Harriet Bulkeley,
Lindsay Mai, Kes McCormick and
Yuliya Voytenko Palgan

Routledge
Taylor & Francis Group

LONDON AND NEW YORK

First published 2018
by Routledge
2 Park Square, Milton Park, Abingdon, Oxon OX14 4RN

and by Routledge
711 Third Avenue, New York, NY 10017

Routledge is an imprint of the Taylor & Francis Group, an informa business

British Library Cataloguing-in-Publication Data
A catalogue record for this book is available from the British Library

Library of Congress Cataloging-in-Publication Data
A catalog record for this book has been requested

ISBN: 978-1-138-71472-4 (hbk)
ISBN: 978-1-138-71477-9 (pbk)
ISBN: 978-1-315-23064-1 (ebk)

Typeset in Bembo and Stone Sans
by Florence Production Ltd, Stoodleigh, Devon, UK

CONTENTS

List of illustrations   *vii*
Notes on contributors   *ix*
Acknowledgements   *xiv*

1   Introduction   1
    Simon Marvin, Harriet Bulkeley, Lindsay Mai, Kes McCormick
    and Yuliya Voytenko Palgan

PART I
Design of ULL   **19**

2   Urban living labs: Catalysing low carbon and sustainable
    cities in Europe?   21
    Yuliya Voytenko Palgan, Kes McCormick and James Evans

3   Putting urban experiments into context: Integrating urban
    living labs and city-regional priorities   37
    Mike Hodson, James Evans and Gabriele Schliwa

4   Urban living labs for the smart grid: Experimentation,
    governmentality and urban energy transitions   52
    Anthony M. Levenda

5   Smart city construction: Towards an analytical framework
    for smart urban living labs   74
    Frans Sengers, Philipp Späth and Rob Raven

PART II
Practices of ULL   **89**

6 Intermediation and learning in Stellenbosch's urban
   living lab   91
   Megan Davies and Mark Swilling

7 Bringing urban living labs to communities: Enabling
   processes of transformation   106
   Janice Astbury and Harriet Bulkeley

8 HomeLabs: Domestic living laboratories under conditions
   of austerity   126
   Anna Davies

9 Urban living labs, "smart" innovation and the realities
   of everyday access to energy   147
   Vanesa Castán Broto

PART III
Processes of ULL   **165**

10 15 years and still living: The Basel Pilot Region laboratory
   and Switzerland's pursuit of a 2,000-Watt Society   167
   Gregory Trencher, Achim Geissler and Yasuhiro Yamanaka

11 Agency, space and partnerships: Exploring key dimensions
   of urban living labs in Vancouver, Canada   189
   Sarah Burch, Alexandra Graham and Carrie Mitchell

12 Placing sustainability in communities: Emerging urban
   living labs in China   210
   Lindsay Mai

13 The importance of place for urban transition experiments:
   Understanding the embeddedness of urban living labs   231
   Frank van Steenbergen and Niki Frantzeskaki

14 Conclusions   248
   *Simon Marvin, Harriet Bulkeley, Lindsay Mai, Kes McCormick
   and Yuliya Voytenko Palgan*

Index   *259*

ILLUSTRATIONS

Figures

3.1	A map of the Corridor district	44
8.1	CONSENSUS living laboratory approach	129
8.2	Washing transition framework	142
10.1	Lifecycle of the 2,000WS-BPR	170
10.2	Paradigm polarities between academia and local government	172
10.3	Trajectory of the 2,000WS-BPR	182
12.1	Mangrove seedlings were planted by Green Source volunteers and fenced up from waves and pollution of irresponsible waste dumping	212
12.2	Carbon GSP "Tanpuhui" platform user guide	216
12.3	State-led research facilities and showcase of smart lamp-post in Nansha	221
12.4	Social–technical split and networking divides	224

Tables

1.1	Strategic, civic and grassroots ULL	8
2.1	Basic information about European projects on ULL	23
2.2	Distinctive features of the design of ULL from the literature	25
2.3	ULL studied by selected European projects	30
7.1	Manor House PACT activities organised by thematic strand	113
8.1	Washing HomeLabs research process	131
8.2	Households recruited in the washing HomeLabs	132
8.3	Irish water milestones	134
8.4	HomeLabs washing types	137

9.1 Top-down and bottom-up models of "smartness" for urban
   planning   152
9.2 Comparative analysis of the articulation of discourses of smartness
   in each case study   160
10.1 Overview of key projects in phase two   180
11.1 Key characteristics of three Vancouver-based projects that illustrate
   urban sustainability experimentation   195
12.1 Carbon GSP place-based experiments   218
12.2 Characteristics of ULL typology in China   226

Boxes

10.1 Coloured solar PVs and second life battery energy storage
   pilot project   181
12.1 Snapshot of a grassroots urban living lab in China   212
12.2 Snapshot of a civic urban living lab in China   216
12.3 Snapshot of a strategic urban living lab in China   221

CONTRIBUTORS

Janice Astbury is a Research Associate at Durham University, UK, where she has contributed to the Governance of Urban Sustainability Transitions (GUST) project and worked with the London Borough of Haringey to address energy vulnerability. She is particularly interested in the role of citizens and community organisations in socially just sustainability transitions.

Vanesa Castán Broto is a Professorial Fellow at the Urban Institute and the Department of Geography at the University of Sheffield, UK. Her research focuses on the governance of climate change in urban areas. She is currently leading a British Academy project studying sustainable energy transitions in urban Mozambique.

Harriet Bulkeley is Professor of Geography at Durham University, UK. Her research focuses on the politics of environmental governance. She currently leads the NATURVATION H2020 project, examining innovative nature-based solutions and the politics of urban sustainability. Her recent books include *An Urban Politics of Climate Change* (2015) and *Accomplishing Climate Governance* (2016).

Sarah Burch is Canada Research Chair (Tier 2) in Sustainability Governance and Innovation and an Associate Professor in the Department of Geography and Environmental Management at the University of Waterloo, Canada. Through her research, writing and teaching, she explores transformative responses to climate change at the community scale, innovative strategies for governing sustainability, and the role that small businesses can play in accelerating urban sustainability transitions.

Anna Davies is Professor of Geography, Environment and Society at Trinity College, Dublin, Ireland, where she directs the Environmental Governance

Research Group and is on the Steering Committee for the Trinity Centre for Future Cities. She has published more than 80 peer reviewed books, book chapters and journal articles.

Megan Davies is a PhD candidate based at the Centre for Complex Systems in Transition at Stellenbosch University, South Africa. Her research looks into the spatial and urban governance implications of the utility scale renewable energy programme in South Africa. She completed her MPhil in Sustainable Development at Stellenbosch University in 2016. Her previous qualifications were also awarded through Stellenbosch University – a Postgraduate Diploma in Sustainable Development in 2013 and BA Knowledge Management and Decision Making in 2011.

James Evans is a Professor of Geography at the School of Environment, Education and Development, University of Manchester, UK. His research interest is in urban environmental governance with a focus on how cities learn to become more sustainable. His research projects focus on urban living labs, smart cities and resilience.

Achim Geissler is a Professor for Energy Efficient and Sustainable Building at the University of Applied Sciences and Arts North-western Switzerland (FHNW). He has researched building physics at University of Kassel, Germany, and gained extensive experience in the building industry in Switzerland and the United Kingdom. His publications cover diverse topics concerning the built environment.

Alexandra Graham recently completed her Masters at the University of Waterloo, Canada, where she focused on municipal climate change adaptation in Metro Vancouver. Representing the University of Waterloo, she was a student delegate at the 2015 United Nations Climate Change Conference: COP21. Alexandra has held leadership positions at non-profit organisations that specialise in community resilience, affordable housing and indigenous rights and has an HBA from the Richard Ivey School of Business. She currently works as a Community Developer where she empowers youth, adults and seniors to be city builders and participate in municipal strategic planning.

Niki Frantzeskaki has a PhD in "Dynamics of Sustainability Transitions" from Delft University of Technology and is an Associate Professor on Sustainability Transitions Governance at DRIFT, Erasmus University Rotterdam, The Netherlands. She is working at DRIFT since 2010 where she researches contemporary sustainability transitions and their governance across Europe and in developing countries. She is coordinating research on environmental governance, transition management and urban sustainability transitions by leading and being involved in a portfolio of research projects including ARTS, GATE, GUST, IMPRESSIONS, ENABLE, RESILIENT EUROPE, SUSTAIN and URBES.

Mike Hodson is Senior Research Fellow in the Sustainable Consumption Institute (SCI) at the University of Manchester, UK. His research interests address the governance of urban and regional transitions to more sustainable futures.

Anthony M. Levenda is a Postdoctoral Research Associate at the School for the Future of Innovation in Society at Arizona State University, USA. His research focuses on the political economy and techno-politics of infrastructures. He is currently researching the politics of expertise in smart cities.

Lindsay (Qianqing) Mai is a Post-doctoral Research Associate in the Department of Geography at the University of Durham, UK. She is interested in studying how the process of public policy making and implementation can influence changes in environmental governance. She works in the areas of urban sustainability governance, climate change mitigation with sectoral interventions and urban transformation with social-technical innovations.

Simon Marvin is Director of the Urban Institute and Professor in the Department of Geography at the University of Sheffield, UK. His research interests focus on the interrelations between urban studies and socio-technical networks.

Kes McCormick is an Associate Professor and Assistant Head at the International Institute for Industrial Environmental Economics (IIIEE) at Lund University, Sweden. Broadly speaking, he works in the fields of sustainability and governance. Specifically, he focuses on the implementation of renewable energy technologies, sustainable urban transformation and education for sustainability.

Carrie Mitchell is a specialist in environmental planning in cities. After completing her PhD in geography at the University of Toronto, she worked as a senior program officer at Canada's International Development Research Centre, funding and managing a portfolio of projects on urban poverty and adaptation to climate change in South and Southeast Asia. Dr. Mitchell's current research program focuses on the politics and planning for urban resilience in North America and adaptation planning and urban services delivery in the global South. She is an assistant professor in the School of Planning at the University of Waterloo, Canada.

Yuliya Voytenko Palgan is an Assistant Professor at Lund University, Sweden. She works in the areas of sustainability, urban governance, innovation and sustainable consumption. Her work focuses on strategies for sustainability solutions in new economies: bio-economy, sharing economy and circular economy.

Rob Raven is Professor of Institutions and Societal Transitions at the Innovation Studies Department of Utrecht University, The Netherlands. His interest is in sustainability transitions and socio-technical innovation and his empirical work has covered energy and mobility transition processes in both Europe and Asia.

He published over 50 scientific articles on these topics and co-editor of the book *The Experimental City* (2016). His current research agenda is focused on analysis of transformative change in urban context such as eco-cities and smart cities. A key question is how socio-technical experimentation, institutional change and incumbent urban regimes co-produce the future of cities worldwide.

Gabriele Schliwa is a doctoral researcher in the School of Environment, Education and Development (SEED), University of Manchester, UK. Her research explores the role of citizens in governance innovation. Having previously worked on urban living labs and smart city projects, her current research focuses on the politics of governing through design.

Frans Sengers is a post-doctoral researcher in the Innovation Studies department of Utrecht University, The Netherlands. His current post-doctoral research centres on eco-city and smart city developments, comparing current activities in Europe and China. He previously completed his PhD dissertation on the Prospects for Sustainable Urban Mobility in Thailand. The connecting theme running through his work is experimentation for transformative change in urban contexts, especially in Asian cities.

Philipp Späth is Senior Researcher at the Institute of Environmental Social Sciences and Geography at Freiburg University, Germany. Trained as a geographer and political scientist, he obtained a PhD in "Science and Technology Studies" in 2009. His main research interest lies on urban and multi-level processes of environmental governance, and how promises and fears around terms like "Smart City" interplay, leading to diverse initiatives that co-shape the knowledge politics and the future of cities.

Frank van Steenbergen has an academic background in Sociology (MSc) and Communication Studies (BA) with a focus on intercultural relations and urban sociology. He joined DRIFT in 2010, where he mainly focuses on social and local dynamics within transition studies. Recurrent themes in his research are social inclusion and exclusion, social innovation, urban marginality, local democratic participation and neighbourhood development. Most of his work at DRIFT relates to neighbourhoods or districts in deprived urban areas. He is also actively involved in exploring innovative forms of (urban) governance.

Mark Swilling is a Distinguished Professor in Sustainable Development, Programme Coordinator: Sustainable Development in the School of Public Leadership and Co-Director of the Centre for Complex Systems in Transition at Stellenbosch University, South Africa.

Gregory Trencher is an Associate Professor in the Graduate School of Environmental Studies at Tohoku University, Japan. His research examines collaborative sustainability initiatives between universities, government, industry and citizenry, and policies for advancing building energy efficiency and retrofitting in large cities. Recent publications appear in Sustainability and Environmental Science and Policy.

Yasuhiro Yamanaka is a Professor in the Faculty of Environmental Earth Science at Hokkaido University, Japan. His research encompasses climate change impacts on marine ecosystems using numerical simulations and social science approaches to investigate formal education, regional communities and sustainable tourism in Hokkaido. Several publications are cited by the Intergovernmental Panel on Climate Change.

ACKNOWLEDGEMENTS

A book project is a collective project. There are however a number of thanks we would like to offer in wider acknowledgement of the process that led to the production of this book. First, we would like to thank JPI Urban Europe for funding the Governance of Urban Sustainability Transitions (GUST) project. This funding created the capacity and resource to take a critical and longer term look at the emergence of urban living labs in a comparative project involving Austria, the Netherlands, Sweden and the United Kingdom. For further information on project outputs see www.urbanlivinglabs.net. This book is not the book of the project. Instead the funding allowed us to carefully assemble an international group of scholars to critically review and discuss their work on urban living labs at a workshop. The origin of the book is in that productive and congenial workshop. Second, we would also like to thank Andrew Mould at Routledge for consistently supporting the production of affordable paperback books in the field of urban sociotechnical studies and our editorial assistant Egle Zigaite for ably taking the book through production. Also to three anonymous referees on the original proposal who provided constructive insights and comments that have improved the book. Finally, many thanks to our collaborators in the GUST project and our colleagues in our respective institutions for providing a supportive and constructive context for the book.

Editors, June 2017

1

INTRODUCTION

Simon Marvin, Harriet Bulkeley, Lindsay Mai,
Kes McCormick and Yuliya Voytenko Palgan

1. Introduction

All cities face a pressing challenge – how can they provide economic prosperity
and social cohesion while achieving environmental sustainability? In response,
new collaborations are emerging in the form of "urban living labs" (ULL) – sites
devised to design, test and learn from social and technical innovation in real time.
ULL are proliferating rapidly across cities internationally as one means through
which this might take place. While the notion of ULL is broad and can be
interpreted in multiple ways, at its heart is the idea that urban sites can provide a
learning arena within which the co-creation of innovation can be pursued between
research organisations, public institutions, private sector and community actors
(Liedtke, Welfens, Rohn and Nordmann, 2012). Through the design and
development of ULL, public-, private- and community-based actors are seeking
to deliver innovative and transformative improvements across the urban milieu,
from buildings to green space, transport to energy systems, local food to sustainable
forms of consumption. For their protagonists, ULL are seen not only as a means
through which to gain experience, demonstrate and test ideas, but also as a step
towards developing responses that have the potential to be scaled up across systems
of provision in order to achieve sustainability transitions at a large scale. However,
the extent to which these experimental interventions can address these urban
challenges has yet to be interrogated. There has to date been relatively little critical
analysis of the emergence, practices and consequences of ULL. This book seeks
to address this deficit.

Our aim is to critically analyse and advance understanding of the governance
of sustainability transitions through a focus on the emergence, practices and
consequences of ULL. These initiatives are proliferating rapidly across the globe
as a means through which public and private actors are testing innovations in
buildings, transport and energy systems. Yet despite the experimentation taking

place on the ground, we lack systematic learning and international comparison across urban and national contexts about their impacts and effectiveness. We have limited knowledge on how good practice can be scaled up to achieve the transformative change required. This is a critical research and policy gap that the book will seek to address.

The core question that this book seeks to ask is whether ULL are best understood as either a new, innovative and transformative method of governing sustainability transitions or whether they are an extension of existing techniques and methods of urban governance that may lead to incremental improvement at best and the continuation of urban social inequalities and environmental challenges at worst. In order to do this, we address three questions. First, we ask what is distinctive about the existing landscape of ULL? Here, we provide an empirical overview of the emergence of ULL, examine how this differs from other forms of urban experimentation, map the location of ULL and undertake detailed empirical analysis of ULL in different urban contexts. Second, we consider how to develop a theoretical approach to analyse ULL. Here, we develop a systematic conceptual framework for evaluating the design, practices and processes of ULL to enable the comparative analysis of their potential and limits. Finally, we ask what are the consequences of ULL – do they lead to transformation or is this more about the maintenance of business as usual? Here, we undertake a comparative analysis within the three parts of the book to provide new conceptual and policy insights into the governance of urban sustainability transitions and improve the design and implementation of ULL in order to understand the implications of this mode of urban experimentation.

Developing this critical perspective is vital if we are to further the use of ULL and understand their limits. Observers have warned that the growing interest in ULL is premature. While technical analyses of the effects of ULL have been conducted (Gil-Castineira, Costa-Montenegro, Gonzalez-Castano, López-Bravo, Ojala and Bose, 2011), we currently lack an integrated assessment of their social, political and economic impacts and how these vary across different national, urban and sectoral contexts. Furthermore, there is limited understanding as to how they can effectively facilitate urban sustainability transitions (Evans and Karvonen, 2013; Nevens, Frantzeskaki, Gorissen and Loorbach, 2012). A range of models are currently being deployed, and yet the empirical data on their relative strengths and weaknesses is fragmented and inconclusive. In order to develop the evidence base that can support ULL as a means for achieving sustainability, a systematic and comparative analysis of their design, operation and impacts is required.

To contribute towards this agenda, the book will connect ULL with the governing of sustainability transitions through an internationally comparative approach examining the connections between: the design of ULL; the practices through which they are implemented and managed; and the processes by which they seek to affect urban systems and governance domains. It develops a robust framework for analysing these three dimensions comparatively in different international multilevel governance contexts. We will focus on ULL that seek to deliver sustainability transitions in the building, transport and energy systems of Africa, Asia, Europe and North America to further the evidence base required to develop new research agendas and shape future policy priorities in this area. The rest of the introduction examines the research and societal importance of ULL, the conceptual framework developed for the book and the structure of the argument.

There are two interlinked sets of rationales for a critical and comparative analysis of the emergence of ULL in both academic and societal developments. The first concerns the role of ULL as a specific form of socio-technical innovation and a distinctive type of experimentation in how urban sustainability may be governed. The second concerns the emergence and rapid acceleration of ULL as a distinctive response in both urban research and urban policy priorities. Each of these rationales are explored in further detail below focusing on the specificity of ULL.

2. The governance and dynamics of ULL

It is increasingly recognised that achieving urban sustainability is not just a matter of gathering more data, creating technical fixes or establishing the right institutions. Rather, transitions are required in the ways in which systems of provision and services are designed, organised and delivered in diverse urban contexts. Such transitions encompass new technologies and infrastructures, but also require shifts in markets, practices, policy and culture (Bulkeley, Castán Broto, Hodson and Marvin, 2010; Frantzeskaki and Loorbach, 2010). To *govern* such transitions remains a key challenge for urban policy-makers, planners and practitioners. In response to the complexities and uncertainties involved, new forms of innovation and experimentation are emerging as a means through which governance can be realised (Bulkeley and Castán Broto, 2012; Frantzeskaki, Wittmayer and Loorbach, 2014; Truffer and Coenen, 2012). ULL offer one such form of experimentation. While the notion of ULL is broad and can be interpreted in multiple ways, at its heart is the idea that urban sites can provide a learning arena within which the co-creation of innovation can be pursued between research organisations, public institutions, private sector and community actors (Liedtke et al., 2012). ULL are seen not only as a means through which to gain experience, demonstrate and test ideas, but also as a step towards scaling-up responses in systems of provision that will have improved effectiveness, political traction and public support. In this sense, ULL are not a stand-alone set of interventions, but part of a wider "politics of experimentation" through which the governing of urban sustainability is increasingly taking place (Bulkeley et al., 2016; Evans, Karvonen and Raven, 2016; McGuirk, Bulkeley and Dowling, 2014).

Understanding the means through which ULL are designed, implemented and take effect can usefully draw on the tradition of innovation studies that has informed the development of the concepts of socio-technical transitions and strategic niche management. Yet ULL, like other forms of socio-technical intervention, are more than merely forms of innovation. They constitute a means through which urban sustainability is governed. Forging a critical and constructive dialogue between insights into the development of urban innovations and the nature of urban governance is regarded as a critical next step for research on sustainability transitions. A key starting point for this analysis has been developing our understanding of how and why specific arenas for innovation might come to play a part in the wholesale transformation of urban systems requires an engagement with research on socio-technical transitions (Smith, Voss and John, 2010). This work has examined the role that niches and experiments play in developing transitions in the face of relatively stable regimes (Schot and Geels, 2008). Existing systems tend to be difficult to dislodge because they are stabilised by regimes and lock-in processes that lead to path dependency and "entrapment", constraining alternatives (Grin, Rotmans and Schot, 2010). One means via which the governance of transitions is thought to proceed is through strategic niche management (Kemp, Schot and Hoogma,1998) – whereby governments, or other actors, deliberately seek to establish conditions under which niches for innovation can grow and "breakthrough" existing regime conditions – or "experimentation", a less directed process which seeks to create spaces for innovation and alternatives to be tested and experience gained (Bulkeley and Castán Broto, 2012). The potential of niches to lead to regime transition is thought to depend on growing the social networks, innovation and learning that they establish (Szejnwald Brown and Vergragt, 2008). Emphasis has been placed on the role of niches as protective environments that provide space for the development, testing and failure of novel innovations, and where new networks can be supported and sustained (Smith and Raven, 2012).

Yet despite its central place in the concept of socio-technical transitions, there has been relatively little analysis of how such forms of protection are afforded. Smith and Raven (2012) argue that alongside processes of shielding and nurturing, niches and experiments foster different forms of empowerment – means through which they are able to either "fit and conform" or "stretch and reform" existing regimes. While research has so far tended to focus on questions of the design and production of niches and experiments, analysing these processes requires that more attention is given to dynamics of agency and power – the practices of governance on the ground (Smith and Raven 2012). At the same time, there are growing calls for research to engage with the spatial and political contexts within which transitions evolve, and the processes through which niches and experiments are related to these wider systemic contexts (Coenen, Benneworth and Truffer, 2012; Meadowcroft, 2007; Shove and Walker, 2007). In this regard, understanding the role of city regions in the governance of transition pathways has become a critical area for research and action (Bulkeley, Castán Broto, Hodson and Marvin, 2010; Coenen and Truffer, 2012; Nevens et al., 2012).

In parallel, there has been a growing interest within urban studies on forms of experimentation and specific ULL taking place in cities. In contrast to the system level perspective offered by transitions research, work on ULL has tended to focus on individual cases and their effects in creating new knowledge and forms of intervention in specific contexts (Evans and Karvonen, 2013). This work has highlighted the importance of intermediaries – specific organisations operating between different social interests (and technologies) – to produce outcomes that would not have been possible without their involvement (Evans and Karvonen, 2013; Hodson and Marvin, 2010; Hodson, Marvin and Bulkeley, 2013). Yet despite insights into ULL in this literature, there is a lack of understanding about which factors facilitate or hinder their success (Shipley, 2000; Van Eijndhoven, Frantzeskaki and Loorbach, 2013) and the processes through which ULL come to shape the governance of wider provision systems.

Each of these approaches – the study of socio-technical systems in transition and urban accounts of new forms of sustainability governance and intermediation – can contribute to our understanding of the dynamics and implications of ULL. In order to understand the role of ULL in urban sustainability transitions, we focus in the remainder of the book on their role in governing sustainability transitions. Drawing together insights from the literatures on socio-technical transitions, urban sustainability and infrastructure governance, we suggest that central to any understanding of the work of ULL in the city are three interrelated processes. The *design* of ULL – the purpose and visions that underpin the initiation of ULL, and the institutional and infrastructural arrangements through which it is enacted. The *practice* of ULL – the techniques, mechanisms and every day actions through which the ULL is implemented and maintained on a day-to-day basis. The *processes* of ULL – the means through which ULL are enrolled within and beyond their specific contexts and come to take effect within socio-technical systems, such as learning, translation, scaling up and so forth. We use these entry points to analyse the case studies presented in the book, but also as a means of understanding the emergence of ULL in the urban governance landscape distinguishing the different forms or modes through which they take place.

3. The emergence and multiple modes of ULL

3.1. Emergence of ULL

Our analysis suggests that ULL have emerged as a particular response to three sets of issues: first, the fragmentation of urban sustainability discourse; second, the challenges of generating systemic change in the organisation of urban infrastructure and the built environment; and third, the introduction of new partners and social interest in urban experimentation. We will briefly look at each of these in turn explaining how these shifts have structured a context in which ULL have become a focus for urban policy and research.

First, there is no longer a singular pathway towards an urban sustainability transition (de Jong, Joss, Schraven, Zhan and Weijnen, 2015). Since the global economic crisis of 2008, there is powerful evidence of a fragmentation of urban sustainability discourse and a search for new modes of urbanism. The research and policy community that had been assembled around the urban sustainability discourse has been unbundled and coalesced around multiple and co-existing urban imaginaries. The most significant logics are smart cities, urban resilience and low carbon cities each associated with a particular coalition of social interests, urban exemplars and transition pathways that seek to remake the city. It is apparent that there is no longer, if there ever was, a singular concept of how the environment of the city needs to be reshaped and reprioritised. Instead there are multiple and competing ideas of what environments are strategically important and require action. Smart cities claim to make material flows more efficient and flexible, low carbon cities reprioritise "carbon" flows and the search for alternatives and urban resilience focuses on those resource flows that are strategic to the protection of cities and seeking to ensure continued urban reproduction. In this context with a multiplicity of ideas and imaginaries many of these new logics claim to be concerned with sustainability transitions (Hodson and Marvin, 2017). Yet it is clear there is a diversity and multiplicity of pathways, the relationships between them are not well understood and it is not readily apparent how these different logics can be understood and integrated. *In this context, ULL are less concerned with developing a singular notion of transition and instead about developing a new capacity to experiment with the intersections and interrelationships between smart, resilience and low carbon.*

Second, there is the challenge of generating systemic change in the social and technical organisation of urban infrastructures and built environment. This is in a context in which there is not clarity about the purposes of the transition, a multiplicity of competing transitions pathways each with quite different implications and an absence of capacity and techniques – particularly within municipalities, especially those operating under conditions of austerity governance, to find ways of systemically shaping programmes of experiments. In this context, there has been a search to find new models and entrants that may be able to overcome obduracy and lock in with existing socio-technical systems and develop new capacity to experiment and innovate. Of particular relevance, there has been the shift to examine techniques and modes of experimentation that sit outside conventional urban planning strategies and frameworks and explore the potential relevance of approaches developed in corporate contexts. It is critically important to recognise that before living labs became "urban", the technique itself was utilised in the corporate sector particularly to develop methods for more open and rapid innovations around products and services often linked with digital and mobile technologies (see Almirall and Wareham, 2011; Bilgram, Brem and Voigt, 2008; Chesbrough, 2003). Key to this was the transmutation of Dutch techniques around open innovation and user-led innovation from the research laboratory into corporate services and product development (von Hippel, 1986, 2005). Working on technologies where potential applications could not be anticipated in advanced major companies developed more open-ended and exploratory techniques working in conjunction with users to develop and experiment with the creation of new producers and services. These techniques recognised that users could be more active in the co-production of new applications and services if innovation methods were more open-ended and experimental. Initially then living labs were not a place-based urban response, they were developed in a corporate environment to more efficiently and rapidly prototype services and products with the users themselves included inside the innovation process. *Living laboratory techniques have transmuted from the research laboratory of corporate enterprises into the urban context with subtle but important shifts as the emphasis became increasingly placed-based and embedded in particular contexts.*

Third, the final shift in the search for new modes of exploring transitions has been an enhanced role for new entrants into the urban innovation process itself. There are two sets of social interests who have become much more active in activities and debates around ULL – we do not claim that these are all new but have become intensified due to a set of three shifts. There has been a significant expansion in corporate prioritisation of the urban environment as a site of experimental activity. Following the global financial crisis, a significant number of corporates in the engineering, consulting computing and digital technologies sectors have developed business strategies that have sought to exploit the urban market for urban infrastructural transitions (Paroutis, Bennett and Heracleous, 2014). Growing urbanisation is leading to a series of market assessment of huge growth in demand for smart city solutions, low carbon investments, and urban infrastructures. On this basis, numerous companies have undertaken to identify the market segment for exploitation through new strategies targeted at developing urban products (Luque-Ayala and Marvin, 2015). This has resulted in the active constitution of the city as a market for integrated products and services – this is an experimental process levering off existing relations with municipal authorities, but also through more active forms of experimentation – competition and sponsorship through companies including IBM, Microsoft and Bloomberg. These processes have brought corporate actors more strongly into urban experimentation – often encouraged through European and national innovation agencies constituting urban as export markets. Parallel to the growth of corporate actors we have seen more active roles for university research beyond the traditional disciplines of urban studies – and involvement of science and technology studies in urban innovation activity. New incentives have been created through national and European research and innovation agencies to stimulate involvement of university science and technology in urban experimentation. Alongside these processes of accelerating the role of corporate and municipal actors there has been an expansion in support for community and grassroots innovations. Innovation agencies have been active in stimulating corporate, university and municipal responses through urban experimentation activities in support of national innovation and economic priorities. *ULL have become prioritised within corporate and national innovation contexts as a technique for experimentation with solutions for urban sustainability.*

3.2. Distinctiveness and modes of ULL

These shifts together have stimulated a resurgence of experimentation in the urban context – ideas about co-production, new roles for users and knowledge partners, competitive forms of bidding and experimentation. These have transmuted into the urban context producing a wide range of experimental responses. Among the growing range of forms of urban experimentation, we can identify ULL as being distinguished by three features. First, placed-based embeddedness – ULL are placed in a geographical area – and concerned with undertaking socio-technical experiments in a particular material setting. Second, experimentation and learning – ULL test new technologies, solutions and policies in real world conditions in highly visible ways with an avowed interest in active forms of learning. Third, participation and user involvement – co-design with stakeholders appears in all stages of the ULL approach. The overarching feature is that some form of evaluation and learning underpins the ability of ULL to facilitate formalised learning among participants, and fulfil their vision to act as urban labs or test beds.

Table 1.1 sets out a typology of ULL that informed the development of our empirical research on ULL. Critically the framework is not a fixed taxonomy designed to reflect reality, but is instead a typology that is designed to be a flexible device that is contestable through the frameworks application in an empirical research setting. The typology is tested through a comparative programme of research that reviewed 50 ULL in different parts of Europe.

TABLE 1.1 Strategic, civic and grassroots ULL

Characteristic	Strategic	Civic	Organic
Lead actors	Innovation agencies, national government Corporate business	Municipal/local authorities Higher education and research institutes Local companies SMEs	Civil society Communities, NGOs, residents, etc.
Urban imaginary	Urban as a test-bed that can be replicated or generalised	Urban as a contingent and historically produced context	Urban understood in particular ways by local communities
Primary purpose	National innovation and technological priorities	Urban economic and employment priorities	Community social, economic and environmental
Organisation Form	Competitive – urban selected as a site for experimentation	Developmental – partnerships formed by local actors	Micro/single – multiple forms of community organisation
Funding type	One-off/competitive	Co-funding/ partnership	Improvised
Analogue	National innovation	Urban technology policy	Grassroots innovation

Strategic ULL are characterised by some degree of steering or conditioning by the national state or regional authorities, as well as the involvement of large corporate or private sector partners. Drawing upon the concept of national innovation systems, the analogy here are national technology development programmes developed by state intermediaries such as Innovate in the UK, FFG Austria, RVO in the Netherlands, Vinnova in Sweden and JPI Urban Europe. These programmes are usually configured to undertake forms of state sponsored experimentation with corporate partners designed to test and develop applications, build local capacity and develop an internationally competitive technology sector. These actors have a tendency to consider their national urban contexts as potential sites for the development and implementation of these experimental activities on the basis of using a live test bed and also trying to establish first mover advantage through the application. The forms of experiments are often varied using a variety of mechanisms – often through competitive processes such as competitions where urban sites assemble partnership and local assets to compete for state funding. Rather than small-scale experiments these tend to be larger scale programmes where experimental initiatives are contained within a wider smart or sustainability strategy. Investment tends to be awarded for a specific activity through a lump sum rather than continued funding over a long period. In this sense, the urban is constituted as an experimental test bed to service other national and corporate priorities.

Civic ULL unlike the strategic mode are much more focused on the priorities of municipal governments and civic universities and locally based companies. While there may still be the involvement of national funders and/or corporate entities the priorities for this form of ULL tend to reflect particular urban priorities and concerns – economic performance, employment creation, overcoming constraints within local infrastructures rather than national innovation priorities per se. The primary goals of these approaches therefore tend to have strong contextual and contingent character although to obtain national funding these local priorities would need to be reworked through the passage point of strategic national state priorities. Co-funding is often a common method for these funding patterns that are often based on a partnership that utilises municipal funding and assets, research and/or engagement funding through higher education institutions and may also seek to utilise national and European funds where appropriate. This style can include stand-alone projects as a one-off experiment or the constitution of a platform or intermediary capacity to undertake a programme of experiments in the city over time. The primary objectives tend to be the transfer of research into demonstration, the development of first mover advantage, innovation and economic development and/or accelerated transition within an infrastructure. In all cases, these ULL attempt to embed learning and benefits within the urban context.

The final style of ULL is those concerned with highly contingent and specific contextual issues that are related to the needs and priorities of particular communities and/or neighbourhoods – these are highly diverse – social needs, unemployment, pollution, fuel poverty, etc. The key actors are urban civil society and not for profit groups – including NGOs, charities, grassroots and community organisations who mobilise residents and publics around specific experimental responses. This framework resonates very powerfully with the grassroots innovation literatures (Seyfang and Smith, 2007) in socio-technical and innovation studies. The focus of activity tends to be on infrastructural innovations that can support different economic, social and environmental dimensions of community well-being and development. Capacity is developed to either respond with a single-issue one-off project or the capability to formulate a programme of activities over a sustained period. Budgets tends to be limited – often dependent on competitive bidding into municipal national and or European funding streams with match funding in volunteer time and other resources.

There are three critical points to make about this landscape of ULL. First, each has a particular set of actors that delimit the configuration and orientation of the ULL. To be clear these are not a mutually exclusive list of actors as there is likely to be considerable overlap – the critical point is that the lead actors shape the primary purpose and goals of the ULL. Second, the actors are assembled within particular types of governance arrangements and forms of partnership that themselves delimit and actively configure the vision and intensions of ULL. Finally, it is of course possible to have a co-existence of different forms of ULL within a single urban context although the extent to which there are active connections or disconnections between these would need to be empirically tested.

4. Critical gaps in the literature

The emergence of ULL has been relatively rapid and accelerated through the intertwining of a series of pressures until they have become a central pillar of urban policy and research priorities. Our concern is that this acceleration and normalisation of ULL as a form of urban research and policy development has proceeded much more rapidly than the development of an evidence base against which to assess the implications and consequences of this mode of experimentation. There are three critical gaps that need to be addressed. First, the need for longer term assessment and comparative analysis of ULL both within the European research landscape and beyond as the concept gains purchase in urban priorities in international agencies (e.g. Belmont Foundation). Second, the need to address the normative focus of current work that often focuses on making ULL work more effective rather than developing a wider critical and conceptual understanding of the emergence of the phenomena – is it different and distinctive? Finally, the need to step back and ask why these responses emerge in particular places – are and what its wider implications might be – how does it reshape our understanding of urban transitions in new ways?

These key gaps mean that the focus of the book has to be based on developing an understanding of three sets of issues. First, to examine the reason for the emergence of ULL in terms of who is involved (and excluded), why are they developed and where they are located. Central to this is analysing why ULL are seen as critical to the governance of sustainability transitions and how the experiences reshape our understanding of modes of governance. Second, there is a need for new empirical analysis of the practices of ULL through detailed case studies of the formation, stabilisation and operation of ULL in different urban contexts. This is necessary to examine what is distinctive about the practices of ULL – do they become embedded and systemic or are they ephemeral and temporary? Finally, there is a need to understand the relationships between the emergence and practices of ULL through an analysis of their consequences and implications. Critical to this are examining what types of innovations are developed and what types of transition pathways may emerge from this – is it meant to shape action in the local context or, for instance, to shape transitions that are designed to happen in other urban contexts. There is, therefore, a need to understand what might be the distinctive contribution of ULL compared to other modes of urban socio-technical innovation.

5. Structure of the book

The book advances the theorization of the interface between experimentation, socio-technical transitions and the city, through an internationally comparative empirical analysis of ULL. This works through three core elements. First, through a set of conceptually informed empirical case studies from the Africa, Asia, Europe and North America, including in-depth analyses of initiatives in ULL. Interviews and case studies with municipal staff, private sector representatives, activists and civic communities, technologists, developers and others involved in the making of ULL. Second, within each case there is a critical examination of the discourses and practices associated with ULL, including an evaluation of what new capabilities are being created by whom and with what exclusions, how these are being developed and contested, where this is happening both within and between cities and with what sorts of social and material consequences. Third, rather than seeking to use one existing conceptual or analytical approach, we develop an original framework informed by the body of work on transitions theory and work on urban governance and politics. The risk of such an approach is theoretical eclecticism. To counter this risk, the framework focuses on basic or core concerns within different approaches and the common ground they share. The book is structured in a threefold framework that is focused on the design of ULL – their purposes and orientation – the practices of ULL – their modes of operation and organisation – and the processes of ULL – their wider implications and consequences. The framework is examined in more detail below.

5.1. ULL design

Part I concentrates on understanding the ways in which ULL are being designed and how they vary between urban contexts. The design of ULL includes forming coalitions and intermediary institutions, establishing shared understanding of challenges of sustainability transitions in relation to particular contexts for intervention, identifying the technological interventions to be trialled and agreeing on the governance principles to be followed. Each set of decisions is open to contestation, and the work required in order to embed and establish ULL is considerable. This part explores the following research questions:

- How are the visions, institutions, intermediaries and technologies required to deliver ULL assembled and embedded, and how and why do these vary across urban contexts?
- What principles and modes of governance are being developed through ULL, and how and why do these vary across contexts?
- Why and how does the design of ULL affect the practices they deploy, their effectiveness and legitimacy? What are the implications for their ability to foster sustainability transitions?

In Chapter 2, "Urban living labs: Catalysing low carbon and smart sustainable cities in Europe?", Yuliya Voytenko Palgan, Kes McCormick and James Evans examine how the ULL concept is being operationalised in contemporary urban governance for sustainability and low carbon cities. This is undertaken through the analysis of major on-going ULL projects. Five key ULL characteristics are identified and elaborated: geographical embeddedness, experimentation and learning, participation and user involvement, leadership and ownership and evaluation and refinement. In Chapter 3, "Putting urban experiments in context: Integrating urban living labs and city-regional priorities", Mike Hodson, James Evans and Gabriele Schliwa critically examine how Urban Living Labs and experiments relate to established urban governance structures and priorities. They are centrally concerned with the ways in which ULL and experiments connect to each other within a city-wide governance context, and how this context both shapes, and is shaped by, experimentation. By considering experimentation within its broader urban context, the chapter sheds light on the potential for portfolios of experiments to shape and drive urban transformation, and by extension, the durability of experimentation as a governance strategy. In Chapter 4, "A living lab for the smart grid: Experimentation, governmentality and urban energy transitions", Anthony Levenda focuses in unpacking the metaphor of the ULL in order to understand the epistemologies underpinning urban experimentation, with particular attention to the ways in which ULL are created materially and discursively as particular places for technological demonstration and testing, so called urban test beds. Tracing the development of the smart grid experiment in Austin, he highlights how the project facilitated demonstration and test-bedding approaches with the primary aim of attracting technology companies to the city, and embedding new logics and rationalities for organizing energy practices and conduct. Finally, in Chapter 5, "Smart city construction: Towards an analytical framework for urban living labs", Frans Sengers, Philipp Späth and Rob Raven mobilise insights from the field of sustainability transitions to provide a fresh analytical perspective on this type of on-going smart city experimentation. At the heart of this perspective are the ideas of the city as a complex patchwork of socio-technical systems, the struggle between obdurate stability and transformative change and the notion of experimentation. They develop an analytical division along the lines of materiality, discourses and institutions to provide a fruitful starting point for the analysis of smart ULL and they illustrate this with examples from the Netherlands and Germany.

5.2. ULL practices

Part II focuses on how, by whom and with what impact ULL are put into practice through detailed case studies. The practice of ULL may involve different ways of and techniques for learning, shielding, nurturing, empowering and participating within ULL. These chapters develop an analysis of the everyday work of ULL and their impacts in particular urban contexts in order to: test existing theories of niche innovation and experimentation, identify which practices are most common across different urban contexts and which practices are most and least successful in delivering environmental, social and economic benefits. In particular the chapters examine:

- Who is undertaking niche practices of learning, shielding, nurturing and empowering within ULL, and how effective are these practices?
- How and why do the practices vary between and within ULL; which practices appear to be more or less successful, and why?
- How can more effective practices be designed to develop ULL in different urban contexts?

In Chapter 6, "Intermediation and learning in the Stellenbosch urban living lab", Mark Swilling and Megan Davies examine the foundations of collaborative governance in Stellenbosch, South Africa to highlight the conditions that allowed for the emergence and evolution of a dynamic ULL. A commitment to practicing trans-disciplinary research and pioneering a unique approach to research partnerships that support the conduct of scientific research with rather than for society opened up novel opportunities for innovation and experimentation. The chapter examines the issues involved in breaking away from traditional "extractive" research modes that define researchers as the generators of solutions and societal actors as the consumers of these solutions. In Chapter 7, "Bringing urban living labs to communities: Enabling processes of transformation", Janice Astbury and Harriet Bulkeley examine urban experimentation for sustainability and the role of community organisations and grassroots innovations. Manor House PACT, in London, UK, was a project that sought to develop and test community-based and community-led responses to urban sustainability challenges designed to engage excluded communities. The chapter focuses on the processes through which PACT has sought to extend its influence beyond its boundaries, particularly focusing on the ways in which the constitution of PACT as part of a wider programme of community initiatives has enabled it to circulate, reaching into arenas and how its approach to learning has been central in fostering its transformative potential. In Chapter 8, "HomeLabs: Domestic living laboratories under conditions of austerity", Anna Davies interrogates the experience of co-designing and conducting a suite of HomeLabs experiments in Dublin, Ireland, seeking to provoke transitions to more sustainable household consumption. Developed with a commitment to co-design principles, the HomeLabs embody a research-led, experimental living laboratory approach. Focusing specifically on the experiences and outcomes of the washing HomeLabs for the participating households and the wider governing environment, this chapter reflects on the opportunities and challenges of such techniques for domestic sustainability transitions under conditions of austerity. In Chapter 9, "Urban living labs, low carbon innovation and the realities of every day access to energy", Vanesa Castán Broto investigates the extent to which smart city visions are fit to deliver both environmental sustainability and social justice in contemporary cities. Thinking about the creation of operational and citizen-oriented smart cities will depend on the extent to which smart city visions, and the means deployed to deliver them, respond to the needs of cities and their material constraints. The analysis follows three case studies in Hong Kong, Bangalore and Maputo. The comparative analysis shows, first, the heterogeneity of perspectives that emerge to characterise smart cities, and, second, their realization as an on-going project of urban development which is never really accomplished.

5.3. ULL processes

Part III investigates the processes through which ULL create an impact beyond their immediate domain, which is critical in terms of assessing their potential as a means of governing transitions for sustainability. These chapters test existing theories concerning how niches and experiments are able to reconfigure regimes and explore the dynamics of these processes. These chapters address three sets of issues:

- How, why and by whom are ULL innovations translated into urban socio-technical systems and regimes; to what extent does this vary according to the territorial context?
- Which mechanisms are most effective in supporting the translation and up scaling of ULL into urban governance domains; to what extent does this vary according to urban context?
- How can the design, implementation and practice of ULL be improved to support the impact and effectiveness of ULL, and can distinct pathways for urban sustainability transitions be identified that can be applied to a wide range of urban contexts?

In Chapter 10, "15 years and still living: The Basel Pilot Region laboratory and Switzerland's pursuit of a 2,000-Watt Society", Gregory Trencher, Achim Geissler and Yasuhiro Yamanaka explore the life-cycle dynamics over the course of the formation, gradual development and subsequent evolvement of a long-running ULL in Switzerland. Through assessing this life-cycle process, they construct an important understanding of the potential long-term societal impacts of ULL, and their capacity to trigger broader societal change. All three phases of the ULL are examined, covering initial design, early implementation and restructured implementation. In Chapter 11, "Agency, space and partnerships: Exploring key dimensions of urban living labs in Vancouver, Canada", Sarah Burch, Alex Graham and Carrie Mitchell examine three initiatives for analysis as place-explicit experimental interventions to sustainability transitions. They represent different phases of innovations, ranging from prototyping to commercialisation to fostering changes within established businesses. In Chapter 12, "Placing sustainability in communities: Emerging urban living labs in China", Qianqing Mai examines the characteristics of emerging ULL in China, mapping out the impacts of three different designs and practices of socio-technical innovations. Grassroots activism is accumulating with increasing social innovations being organised by voluntary community leaders. Civic type of community innovations is also coordinated by municipal authorities, universities and government-organised non-governmental organisation, indicating higher mainstreaming capacity across subnational regions. Strategic experimentation of smart urbanism is also discussed concerning technological innovations being systematically rolled out in a state-endorsed new urban district. In Chapter 13, "The importance of place for urban transition experiments: Understanding the embeddedness of urban living labs", Frank van Steenbergen and Niki Frantzeskaki employ the concept of "sense of place" to understand the embeddedness of ULL in a particular socio-spatial context in a relational perspective. The sense of place, as outcomes of experimentation, entails the collection of meanings, beliefs, symbols, values and feelings held by social members in association with a particular locality. The chapter then looks into how experimental methodologies in participatory engagement had been applied in an urban regeneration intervention of Veerkracht Carnisse in Rotterdam to empower local communities and to make the area more sustainable and resilient.

Finally, in the conclusion the editors reflect on the objectives of the book through the specific questions of each part and then set out a future research and policy agenda.

References

Almirall, E. and Wareham, J. (2011). 'Living labs: Arbiters of mid- and ground-level innovation'. *Technology Analysis and Strategic Management, 23*(1), 87–102.

Bilgram, V., Brem, A. and Voigt, K.-I. (2008). 'User-centric innovations in new product development'. *International Journal of Innovation Management, 12*(3), 419–458.

Bulkeley, H. and V. Castán Broto. 2012. 'Government by experiment? Global cities and the governing of climate change'. *Transactions of the Institute of British Geographers, 38*(3), 361–375.

Bulkeley, H., Castán Broto, V., Hodson, M. and Marvin, S. (2010). *Cities and low carbon transitions*. New York: Routledge.

Bulkeley, H., Coenen, L., Frantzeskaki, N., Hartmann, C., Kronsell, A., Mai, L. . . . and Voytenko Palgan, Y. (2016). 'Urban living labs: Governing urban sustainability transitions'. *Current Opinion in Environmental Sustainability, 22*, 13–17.

Chesbrough, H.W. (2003). *Open innovation: The new imperative for creating and profiting from technology*. Boston, MA: Harvard Business School Press.

Coenen, L., Benneworth, P. and Truffer, B. (2012). 'Toward a spatial perspective on sustainability transitions'. *Research Policy, 41*(6), 968–979.

Coenen, L. and Truffer, B. (2012). 'Spaces and scales of sustainability transitions: Geographical contributions to an emerging research and policy field'. *European Planning Studies, 20*(3), 367–374.

de Jong, M., Joss, S., Schraven, D., Zhan, C. and Weijnen, M. (2015). 'Sustainable-smart-resilient-low carbon-eco-knowledge cities; making sense of a multitude of concepts promoting sustainable urbanization'. *Journal of Cleaner Production, 109*, 25–38.

Evans, J. and Karvonen, A. (2013). 'Governance of urban sustainability transitions: Advancing the role of living laboratories'. *International Journal of Urban and Regional Research*. doi:10.1111/1468-2427.12077

Evans, J., Karvonen, A. and Raven, R. (2016). *The experimental city*. London: Routledge.

Frantzeskaki, N. and Loorbach, D. (2010). 'Towards governing infrasystem transitions, reinforcing lock-in or facilitating change?' *Technological Forecasting and Social Change, 77*(8), 1292–1301.

Frantzeskaki, N., Wittmayer, J. and Loorbach, D. (2014). 'The role of partnerships in "Realizing" urban sustainability in Rotterdam's city ports area, the Netherlands'. *Journal of Cleaner Production, 65*, 406–417.

Gil-Castineira, F., Costa-Montenegro, E., Gonzalez-Castano, F.J., López-Bravo, C., Ojala, T. and Bose, R. (2011). 'Experiences inside the ubiquitous Oulu Smart City'. *Computer, 44*(6), 48–55.

Grin, J., Rotmans, J. and J. Schot. (2010). *Transitions to sustainable development: New directions in the study of long term transformative change*. London: Routledge.

Hodson, M. and Marvin, S. (2010). 'Can cities shape socio-technical transitions and how would we know if they were?' *Research Policy, 4*(39), 477–485.

Hodson, M. and Marvin, S. (2017). 'Intensifying or transforming sustainable cities? Fragmented logics of urban environmentalism'. *Local Environment, 22*(1), 8–22.

Hodson, M., Marvin, S. and Bulkeley, H. (2013). 'The intermediary organisation of low carbon cities: A comparative analysis of transitions in Greater London and Greater Manchester'. *Urban Studies, 50*(7), 1401–1420.

Kemp, R., Schot, J. and Hoogma, R. (1998). 'Regime shifts to sustainability through processes of niche formation: The approach of strategic niche management'. *Technology Analysis and Strategic Management, 10*(2), 175–198.

Liedtke, C., Welfens, M., Rohn, H. and Nordmann, J. (2012). 'LIVING LAB: User-driven innovation for sustainability'. *International Journal of Sustainability in Higher Education, 13*(2), 106–118.

Luque-Ayala, A. and Marvin, S. (2015). 'Developing a critical understanding of smart urbanism?' *Urban Studies, 52*, 2105–2116.

McGuirk, P., Bulkeley, H. and Dowling, R. (2014). 'Practices, programs and projects of urban carbon governance'. Perspectives from the Australian city. *Geoforum, 52*, 137–147.

Meadowcroft, J. (2007). 'Who is in charge here? Governance for sustainable development in a complex world'. *Journal of Environmental Policy and Planning, 9*(3–4), 299–314.

Nevens, F., Frantzeskaki, N., Gorissen, L. and Loorbach, D. (2012). 'Urban transition labs: Co-creating transformative action for sustainable cities'. *Journal of Cleaner Production*, *50*(1), 111–122.

Paroutis, S., Bennett, M. and Heracleous, L. (2014). 'A strategic view on smart city technology: The case of IBM Smarter Cities during a recession'. *Technological Forecasting and Social Change*, *89*, 262–272.

Schot, J. and Geels, F.W. (2008). 'Strategic niche management and sustainable innovation journeys: Theory, findings, research agenda, and policy'. *Technology Analysis and Strategic Management*, *20*(5), 537–554.

Seyfang, G. and Smith, A. (2007). 'Grassroots innovations for sustainable development: Towards a new research and policy agenda'. *Environmental Politics*, *16*(4), 584–603.

Shipley, R. (2000). 'The origin and development of vision and visioning in planning'. *International Planning Studies*, *5* (2), 227–238.

Shove, E. and Walker, G. (2007). 'CAUTION! Transitions ahead: Politics, practice and sustainable transition management'. *Environment and Planning A*, *39*, 763–770.

Smith, A. and Raven, R. (2012). 'What is protective space? Reconsidering niches in transitions to sustainability'. *Research Policy*, *41*(6), 1025–1036.

Smith, A., Voss, J.-P. and John, G. (2010). 'Innovation studies and sustainability transitions: The allure of the multi-level perspective and its challenges'. *Research Policy*, *39*(4), 435–448.

Szejnwald Brown, H. and Vergragt, P.J. (2008). 'Bounded socio-technical experiments as agents of systemic change: The case of a zero-energy residential building'. *Technological Forecasting and Social Change*, *75*, 107–130.

Truffer, B. and Coenen, L. (2012). 'Environmental innovation and sustainability transitions in regional studies'. *Regional Studies: The Journal of Regional Studies Association*, *46*(1), 1–21.

Van Eijndhoven, J., Frantzeskaki, N. and Loorbach, D. (2013). 'Connecting long and short-term via envisioning in transition arenas, how envisioning connects urban development and water issues in the city of Rotterdam, The Netherlands'. In J. Edelenbos, N. Bressers and P. Scholten (Eds.), *Connective capacity in water governance* (pp. 172–190). Farnham: Ashgate.

Von Hippel, E. (1986). 'Lead users: A source of novel product concepts'. *Management Science*, *32*, 791–805.

Von Hippel, E. (2005). *Democratizing innovation*. Cambridge, MA, and London: MIT Press.

PART I
Design of ULL

2

URBAN LIVING LABS

Catalysing low carbon and sustainable cities in Europe?

Yuliya Voytenko Palgan, Kes McCormick and James Evans

1. Introduction

Different forms of urban governance are being developed and tested in European cities. Urban living labs (ULL) constitute a form of experimental governance whereby urban stakeholders develop and test new technologies and ways of living to address the challenges of climate change and urban sustainability (Bulkeley and Castán Broto, 2013). For cities trying to position themselves as innovation leaders in the race to decarbonise and become sustainable, ULL are both high-profile statements of intent and increasingly essential vehicles to secure funding for sustainable urban development (Schliwa, 2013). For funding bodies and governments, they offer a way to encourage cities to adopt innovative solutions. That said, their recent and rapid proliferation has prompted debate in academic and professional circles concerning whether a living lab approach can help catalyse urban sustainability and low carbon transitions.

Living labs for sustainability, low carbon and smart cities that have emerged in Europe have different goals and ways of working, they are initiated by various actors, and they form different types of partnerships. There is clearly no uniform definition of living labs (Hillgren, 2013; Schliwa, 2013; Ståhlbröst, 2008). Some scholars and organisations define them as partnerships between sectors (often between public, private and people) (Börjeson, 2008; EC, 2015; European Network of Living Labs [ENoLL], 2015; Rösch and Kaltschmitt, 1999) where universities play a key role (Evans and Karvonen, 2010), while others look at living labs more in the light of pilot and demonstration projects, which function as supportive tools for private actors and industry helping them commercialise their services, products and technology (Hellström Reimer, McCormick, Nilsson and Arsenault., 2012; Kommonen and Botero, 2013). Living labs can be considered both as an *arena* (i.e.

geographically or institutionally bounded spaces), and as an *approach* for intentional collaborative experimentation between researchers, citizens, companies, and local governments (Schliwa, 2013).

At the same time, while many living labs are emerging, there is no clear understanding of the ultimate role ULL can or should play in urban governance, whether they represent a completely new phenomenon or if they are replacing other forms of participation, collaboration, experimentation, learning and governance in cities. There is a need to clarify what makes the ULL approach attractive and novel, including why funding agencies are investing in exploring its usefulness and why local collaborations are trying to operationalise the concept in real-life settings, and the potential impacts of ULL and their ability to catalyse urban sustainability and low carbon cities. Therefore, the aim of this chapter is to contribute to knowledge on how ULL are being designed and operationalised in contemporary urban governance for sustainability.

The basis of this chapter is a review of academic publications, policy and grey literature, and current or recently completed research projects on urban governance and living labs in Europe. It is supported by a snapshot case study analysis of five research projects funded by the Joint Programming Initiative (JPI) Urban Europe, which are designed to explore or apply ULL methodology with the purpose to contribute to sustainable cities. These snapshots are five of the 20 European projects funded by the JPI Urban Europe within its two targeted calls in 2012 and 2013 (Table 2.1). They are selected as they use the terminology "living labs", "urban living labs", "urban labs" and/or "city labs" to study, explore, test or apply living labs methodology either to an existing urban infrastructure or by establishing new ULL, and study or create ULL with an explicit objective to tackle urban sustainability and decarbonisation challenges. The chapter concludes by assessing the main claims and assumptions identified in the existing literature and highlights areas for future study.

2. ULL through the lens of governance

ULL constitute a mode of governance that promises to deliver valuable outcomes by bringing relevant stakeholders together to address challenges and produce solutions in real world settings. But while ULL are proliferating, their design, impacts and implications for urban governance remain largely unexamined. The emergence of ULL can be situated within a broader diversification of governance over the last 25 years. In response to increasingly restricted municipal funding, municipalities have turned to partnership-based modes of governance that bring public bodies, universities, government, and industry together to address specific sectoral and spatial challenges (Percy, 2008). In this sense, ULL are continuous with urban development approaches from the 1990s onwards that have been characterised by partnerships and area-based initiatives that focus on discrete interventions in particular places.

TABLE 2.1 Basic information about European projects on ULL

No.	Project acronym	Project aim	ULL relevance
1	APRILab: Action-oriented research on planning, regulation and investment dilemmas in a living lab experience	To explore political dilemmas that constrain innovation in cities	The living lab approach is used as a methodology to study existing urban structures and processes
2	CASUAL: Co-creating attractive and sustainable urban areas and lifestyles	To explore how to promote sustainable living and consumption in cities via stakeholder engagement	The usefulness of a living lab approach is explored by creating, managing and studying two ULL
3	Green/blue cities; green/blue infrastructure for sustainable, attractive cities	To develop knowledge and tools to use green and blue infrastructure in New Kiruna City to handle storm water	The living lab approach is used to bring citizens, practitioners, decision makers and researchers together, and jointly develop innovative solutions for managing storm water
4	SubUrbanLab: Social uplifting and modernization of suburban areas with the urban living lab approach	To examine how suburbs can be modernised and socially uplifted to make them more attractive, sustainable and economically viable	The living lab approach is used to integrate people into the development, planning and implementation of actions to modernise the suburbs
5	URB@EXP: Towards new forms of urban governance and city development: Learning from URBan EXPeriments with living labs and city labs	To develop guidelines on the types of problems for which ULL are most suited and how they can be best integrated into formal local government organisations	The usefulness of a living lab approach is explored by reviewing experiences of urban labs, and through action research in urban labs in five European cities

But while clearly fitting in to a longer lineage of urban governance, ULL are also suggestive of something new. Living labs have their origins in the corporate world of ICT research and development, as an innovation approach to develop and refine products and services in real world settings. The underlying logic here is to develop innovations in real settings, such as buildings, with intended users in order to make them more likely to be usable in practice, and hence accelerate their adoption (Evans and Karvonen, 2010). ULL have proliferated very quickly in Europe due to substantial financial and policy backing from the European Commission, seeking to address the so-called European paradox whereby leadership in innovation and sustainability fails to translate into commercial success (Veeckman, Schuurman, Leminen and Westerlund, 2013). Their potential to catalyse rapid uptake of innovations has increasingly positioned them as key drivers for low carbon and sustainability transitions, where new solutions and discrete successes have struggled to be rolled out or up-scaled. For municipalities trying to establish themselves as innovation leaders in the field of sustainability and smart technologies, ULL are high profile statements of intent and effective vehicles with which to secure national and European funding. For funding bodies and governments, specifying ULL projects offers a way to encourage risk-averse authorities into adopting innovative urban solutions without having to engage in the more politically fraught processes of structural or policy reform (Voytenko, McCormick, Evans and Schliwa, 2016).

Climate change has been among the forces driving the adoption of ULL by local policy makers to cultivate "new techniques of governance" for urban sustainability (Hodson and Marvin, 2007). This trend epitomises the turn to experimental approaches to governing climate adaptation (Bulkeley and Castán Broto, 2013). Climate experiments represent the practical dimension of adaptation – they are what happens in practice when policy makers, researchers, businesses and communities are charged with finding new paths (Evans, 2011). The appeal of experimentation (as shown above in the ULL case) is that testing out new technologies and policies under real world conditions in highly visible ways can prompt radical social and technical transformation (Evans and Karvonen, 2010). ULL represent a specific form of experimentation, whereby processes of innovation and learning are formalised (Evans and Karvonen, 2010), and it is this that sets ULL apart from more general policy experiments or innovation niches (Bulkeley and Castán Broto, 2013).

Another defining characteristic of ULL design is that they territorialise urban innovation at a manageable scale, but within this there is considerable variation. Juujärvi and Pesso (2013) define ULL as physical regions "in which different stakeholders form public–private–people partnerships of public agencies, firms, universities, and users collaborate to create, prototype, validate, and test new technologies, services, products, and systems in real-life contexts". They suggest that ULL are characterised by a focus on "urban" or "civic" innovation, which strengthens the public elements of urban innovation. Furthermore, an emerging body of work seeks to understand ULL as Urban Transition Labs, drawing conceptually on the field of transition management to suggest how they can co-create pathways to wider urban change (Nevens, Frantzeskaki, Gorissen and Loorbach, 2012). Drawing on the insights of transition management, this work suggests that the degree to which ULL are able to stimulate broader changes beyond their institutional and spatial boundaries is directly related to the design of ULL partnerships, whereby the exact composition and structure determine which actors are included and the collective rules of experimentation.

Representing an experimental mode of governing, ULL aim at transforming urban governance as they enact and test sustainable innovations by providing platforms for knowledge co-production with multiple stakeholders and users (Baccarne, Schuurman, Mechant and De Marez, 2014). The ULL approach is based on the much-vaunted quadruple helix model of partnership whereby government, industry, the public and academia work together to generate innovative solutions. The knowledge partners often represented by leaders from universities and research institutes appear to be driving and delivering the formalised processes of learning in ULL through evaluation and refinement of ULL goals, visions and practices (König and Evans, 2013; Trencher, Bai, Evans, McCormick and Yarime, 2014), forming another distinct feature that sets ULL apart from other experimental modes of governing. Table 2.2 summarises the key ULL design features discussed above that distinguish ULL from other urban development approaches. These are structured around four questions: "how?", "where?", "what?" and "who?"

Within the rapid uptake of ULL, it is possible to find examples of both techno-centric and socially driven forms of innovation. While many ULL "offer a potential to achieve a low carbon economy by developing innovative energy solutions, stimulating greater cross-disciplinary research in universities and enhancing the ties between institutions that create knowledge and those that use it" (Evans and Karvonen, 2013), not all of them are designed with a low carbon rationale in mind. ULL can also be oriented at promoting economic growth or enhancing social cohesion. In terms of impacts and implications for urban governance, research highlights the risk that overly techno-centric ULL fail to produce innovation or learning and can be easily co-opted by dominant economic interests. In their study of the Clean Urban Transport Europe Programme that trialled green transport solutions in major European cities, Hodson and Marvin (2009) argue that these projects are little more than demonstrations of existing technologies and services, and that they did not engage local populations or context. Work conducted on the Oxford Road Corridor in Manchester, UK, has highlighted similar challenges concerning how to achieve social inclusion in ULL, and the

TABLE 2.2 Distinctive features of the design of ULL from the literature

Key question	ULL design feature	Explanation
How are ULL designed?	Visibility	As highly visible innovations thus potentially triggering transformation
Where are ULL designed?	Territorialisation	In urban domain at a manageable scale
What form do ULL take when designed?	Experimentation and learning	A specific form of experimentation, where innovation and learning are formalised
Who is involved in ULL design?	Participation and leadership	Quadruple helix actors with knowledge partners delivering formalised learning

de-politicisation of urban governance that corporate-led partnerships and scientific modes of governance threaten (Evans, 2011; Evans and Karvonen, 2014).

The migration of the living lab methodology into the urban sphere raises a series of challenges, relating to the dangers of replacing messy political processes of urban development with innovation methodologies (Evans et al., forthcoming). An example of Nexthamburg ULL in Germany showed a remarkable process of user involvement with a very high degree of citizen co-creation of the urban vision for Hamburg, but little adoption and implementation of the suggested ideas by the municipality afterwards due to its low involvement in the ULL (Menny, 2016). In contrast, the Malmö Innovation Platform in Malmö, Sweden, which applies a living labs approach, integrates social engagement directly into technical and economic objectives (McCormick and Kiss, 2015). This highlights the need to understand the design of ULL including their local context, the primary motivations for setting up an ULL and the role of actors owning and leading ULL, and collaborating in them. The remainder of this chapter examines the design, establishment and goals of cutting-edge ULL initiatives within their broader urban context.

3. Characteristics of ULL

When analysing the insights from the theory and literature (Table 2.2), and comparing them with the findings from empirical examples of ULL projects and cases in Europe (see also the following sections), five key characteristics of ULL can be identified: geographical embeddedness, experimentation and learning, participation and user involvement, leadership and ownership and evaluation and refinement. Each of these is briefly discussed here. First, ULL are *placed in a geographical area* – they are predominately not virtual platforms, although they may utilise online tools. ULL represent ecosystems of open "urban" or "civic" innovation, and they are situated in a real urban context where the process in focus is taking place. This may be a region, an agglomeration, a city, a district or neighbourhood, a road or corridor or a building. There are many possible urban configurations that can host ULL, but the area is normally clearly defined and has a manageable scale. Second, ULL represent a specific form of *experimentation*, whereby processes of innovation and *learning* are formalised – unlike policy experiments and innovation niches. ULL test new technologies, solutions and policies in real world conditions in highly visible ways, which can prompt radical social and technical transformation. An important component of this experimentation is the co-production of knowledge and ideas with the users. By placing user-centred experimentation at their heart, ULL are open to unexpected discoveries and learning that originates from the users. ULL also apply user-centred experimentation to achieve a wider learning experience and exercise innovative forms of urban governance based on actor participation.

Logically the third characteristic of ULL enters here, which is *participation and user involvement*. Participation and co-design with stakeholders such as residents and users is at the core and appears in all stages of the ULL approach – from identifying stakeholder needs, deciding upon ULL goals and visions, planning and designing to developing, implementing, evaluating ULL actions and updating ULL ambitions. The interaction process and methods should be differentiated accommodating the background and interests of different stakeholder groups. ULL represent research and innovation processes within a public–private–people partnership (bringing together citizens, practitioners, decision makers and researchers). This also means that research organisations and government funding bodies become more actively engaged in sustainable urban development to help address gaps in knowledge and finance. An important practical challenge for many ULL lies in how to achieve the inclusion of all key relevant stakeholders (both active and passive), account for their interests and thus re-politicise this new form of urban governance that corporate-led partnerships and scientific modes of governance might threaten.

This argument feeds into the discussion of *leadership and ownership* of ULL, the fourth key characteristic identified. The literature suggests that the ability of ULL to contribute to urban sustainability and low carbon transitions depends on how they are designed and executed in practice, and it appears from the case studies that having a clear leader or owner is crucial for ULL. There is an important coordination and management role for ULL to be effective, although a delicate balance exists between steering and controlling. ULL need to remain flexible for different stakeholders to engage in their development and direction. Finally, *evaluation* of the actions and impacts of ULL is important to feed back the results, and revisit and refine the goals and visions over time. Evaluation underpins the ability of ULL to facilitate formalised learning among the participants, and is in most cases performed by ULL knowledge partners such as research institutes or universities. Continuous assessment and feedback within ULL is highlighted strongly in the case studies, but in reality it appears that these objectives are not always met by the ULL.

4. Similarities and differences in the design of ULL

The partners and partner countries in the case studies demonstrate a strong bias towards Northern Europe and Scandinavia, which reflects a legacy of expertise in living lab approaches in these places. Finland in particular has played a key role in developing the living lab methodology over the past 20 years as a broad innovation tool for product and service development. The partners and partner countries also reflect the geographical distribution of expertise in sustainable urban planning, with Sweden and the Netherlands in particular strongly represented. To some extent this reflects the relatively high contributions of the Nordic and northern European countries to the JPI funding. But this in turn indicates the attraction of the ULL approach to these countries, which resonate with a broader trend towards the integration of innovation, sustainability and urban agendas, and governance cultures that value partnerships.

When analysing ULL used by the selected projects, four key topics are found relevant to discuss similarities and differences of ULL design:

- the ways in which the ULL approach is operationalised based on different goals;
- the type of partnership in ULL and the role of knowledge partners in the ULL;
- the types of challenges and topics addressed by different ULL; and
- the role of sustainability, environment, smart and low carbon agendas in ULL.

First, all projects analysed in this chapter are driven by the need to tackle existing urban challenges. They, however, have different goals and thus view the usefulness of the ULL concept and approach for their work in different ways. Some projects create and manage ULL (SubUrbanLab, CASUAL, URB@EXP), while others analyse the suitability of the ULL approach (APRILab, URB@EXP, Green/Blue Cities) to address the pressing dilemmas in cities. The URB@EXP project studies existing ULL and experimentation activities in the partner cities to develop guidelines on the types of problems for which ULL are most suited and how they can be best integrated into formal local government organisations while the CASUAL and APRILab projects investigate the broader potential of ULL for urban governance and planning. URB@EXP defines ULL somewhat more narrowly than other projects by stressing that it is primarily local governments, who engage with other stakeholders to solve urban development problems. However, there is a diversity of types of ULL and while local governments are a common actor to be involved, they do not always lead or coordinate. URB@EXP has therefore focused on a particular design of ULL in which local governments play the leading role.

Second, all projects are led by knowledge partners – mainly research institutions, which play central roles by driving the case study selection and defining visions for ULL and their applicability, and also designing and setting up ULL (e.g. CASUAL, URB@EXP). In three out of five preliminary selected cases of urban areas in the APRILab project, the university campuses are located within the area (Aalto University in Espoo City and Aalborg University, both in South Harbour of Copenhagen, and in the Aalborg East area in Aalborg). Such a strong role of research institutions can be explained by the fact that all studied projects are funded by JPI Urban Europe, which in its two first calls has heavily supported research (both basic and applied) and only to a certain extent innovation (essentially the funds provided by local funding bodies within this programme are insufficient to fund pilot or demonstration initiatives). This empirical finding also reinforces the finding from the literature concerning the importance of knowledge partners in ULL as one of the defining features of ULL design (Table 2.2).

This also may be one of the reasons why the private sector does not seem to be heavily represented in the design of ULL used in the projects (only two projects have SME partners in their consortium: an exhibition and meeting space Färgfabriken in CASUAL, and a foresight and design studio Pantopicon in URB@EXP). Another explanation for the somewhat low representation of the private sector may be linked to the fact that most of the projects have started quite recently, and may have not yet defined the partnerships for their ULL or have not disseminated the information about such partnerships. While all the projects demonstrate their strong connections to local governments and city actors, only two have municipal partners as (funded) members in the consortium (SubUrbanLab and URB@EXP).

Third, the projects studied and their selected ULL address a great variety of topics that are driven by different urban sustainability challenges (see Table 2.3). These include:

- urban planning (e.g. a need for densification, zoning and development of mix-use areas, development of public transportation in APRILab, low attractiveness of disadvantaged areas in APRILab and SubUrbanLab, planning for green space and nature in the city in URB@EXP);
- social development (e.g. segregation, unemployment, low level of education among inhabitants in APRILab);
- economic growth (e.g. decreasing investments in APRILab, a need for business development in APRILab and SubUrbanLab);
- environmental sustainability (e.g. handling storm water in Green/Blue Cities); and
- consumption and lifestyles (e.g. changed living preferences of households in APRILab, questions of housing, mobility and sustainable lifestyles in CASUAL and URB@EXP).

Fourth, when discussing the role of sustainability, environment, smart and low carbon agendas in relation to ULL, only one project analysed in this chapter – the Green/Blue Cities – has a clear rationale to address the challenge of climate change. In the APRILab project, the low carbon agenda is explicitly mentioned only in the case of Copenhagen as the city aims to become sustainable and zero carbon by 2025 (Hansen, Savini, Wallin and Mäntysalo, 2013). In the SubUrbanLab project, the topics of ULL in Finland have a slightly stronger environment and low carbon perspective as two of them seek to decrease energy consumption and promote the use of sustainable energy. The smart city agenda seems to be implicitly present only in one ULL case (ULL "Energetic cooperation" in Peltosaari district of Riimäki within SubUrbanLab project), which seeks to decrease electricity consumption by providing smart meters to residents. Other studied ULL are not biased towards smart city technologies.

When exploring how other environmental challenges are considered by the studied projects, in most APRILab cases they do not have high priority, and they are often presented as "add on" dilemmas. Only the IJburg project in Amsterdam contains a strong environmental component, which is linked to water–land planning

TABLE 2.3 ULL studied by selected European projects

No.	Project	ULL case study	Country	What is explored/achieved through the ULL case study
1	APRILab	An urban development project T3 in Espoo	Finland	The ULL approach is used to explore solutions to existing challenges: urban sprawl, a need for densification and creating a mixed-use area, development of public transportation (metro, light rails, buses)
		The South Harbour neighbourhood in Copenhagen	Denmark	The ULL approach is used to explore solutions to existing challenges: low attractiveness of disadvantaged areas, inequality in urban development and investments
		Aalborg East neighbourhood in Aalborg	Denmark	The ULL approach is used to explore solutions to existing challenges: segregation, mono-functional areas with big distances, unemployment, many living on welfare, low level of education among inhabitants
		Post-suburban development IJburg in Amsterdam	The Netherlands	The ULL approach is used to explore solutions to existing challenges in IJburg: decreasing investments, stringent environmental regulations, environmentally sound water–land planning and accounting for a protected ecosystem of the IJmeer, changed living preferences of households
		Overamstel project area in Amsterdam	The Netherlands	The ULL approach is used to explore solutions to existing challenges: urban intensification and integration of the area to the urban structure of the city, zoning and land use mix, business improvement, noise environmental zoning
2	CASUAL	A neighbourhood living lab in Vienna	Austria	The ULL on transport and mobility is created and managed to generate innovative ideas and scenarios for sustainable urban development
		A neighbourhood living lab in Stockholm	Sweden	The ULL on housing and lifestyle is created and managed to generate innovative ideas and scenarios for sustainable urban development

continued . . .

TABLE 2.3 Continued

No.	Project	ULL case study	Country	What is explored/achieved through the ULL case study
		ULL Färgfabriken in Stockholm	Sweden	An existing ULL, which is an exhibition and a meeting space for art, architecture and urban development, is a partner in the project and is studied as a reference
3	Green/blue cities	An international ULL in Kiruna	Sweden	The ULL approach is used to bring together citizens, practitioners, decision makers, and researchers, to jointly develop innovative solutions on green/blue infrastructure
		A national ULL	Austria	
		A national ULL	The Netherlands	
4	SubUrbanLab	ULL "Shape your world" in Alby suburb of Botkyrka	Sweden	Creating an ULL to promote youth and urban gardening, and modernisation and social uplifting of suburbs
		ULL "New light on Alby Hill" on Alby hill of Botkyrka	Sweden	Creating an ULL to experiment with LED lighting to make an area more secure and attractive
		ULL "Vacant space Alby" in central Alby of Botkyrka	Sweden	Creating an ULL to use abandoned space for activities by residents
		ULL "Energetic cooperation" in Peltosaari district of Riimäki	Finland	Creating an ULL to decrease electricity consumption by providing smart meters to residents
		ULL "Sustainable decisions" in Peltosaari district of Riimäki	Finland	Creating an ULL to raise awareness about sustainable energy for decision makers
		ULL "Together more" in Peltosaari district of Riimäki	Finland	Creating an ULL to offer new house functions in a partially empty community building
5	URB@EXP	The Maastricht-LAB in Maastricht	The Netherlands	Studying an existing ULL, which started in 2011 and implemented eight projects to tackle complex urban challenges (e.g. dealing with the issue of vacant property in the city)
		Stadslab2050 in Antwerp	Belgium	Studying an existing ULL, which started in 2013 and developed 15 project ideas on green space and

continued . . .

TABLE 2.3 Continued

No.	Project	ULL case study	Country	What is explored/achieved through the ULL case study
				nature in the city, and sustainable living and renovation; these were subsequently designed and implemented; new thematic challenges are considered
		City of Malmö	Sweden	The City of Malmö has been strongly involved in living labs hosted at MEDEA/K3, Malmö University. The City of Malmö does not currently run its own ULL, but it is interested in exploring options to establish an ULL for new forms of urban governance and city development
		City of Graz	Austria	The City of Graz aims to gather the latest experience from Europe on how to involve citizens in city development to make urban areas more suitable to their needs and prevent social problems. Living labs are an integral part of the Smart City Graz Action Plan 2020
		City of Leoben	Austria	The City of Leoben sees ULL as necessary for a broader view on stakeholder needs and perceptions, and that it can help close the gap between politics and citizenship; the clarification on who will be the best candidates for ULL is needed

and the proximity of the protected ecosystem of the IJmeer (Hansen et al., 2013). The development of public transportation (light rail and metro) is central to the T3 project in Espoo city (APRILab), which will have implications for reducing greenhouse (GHG) emissions; however, it does not appear as a primary goal in the development of the area. The focus of CASUAL is clearly on sustainable and environmentally conscious consumption as it aims to explore how to promote sustainable living.

Although urban climate governance is characterised by experimental approaches, the trends indicate that not all ULL are designed with a low carbon rationale in mind, but tend to be oriented towards more mainstream urban development challenges associated with promoting economic growth or enhancing social cohesion. The topics often include modernisation and social uplifting of districts or regions, challenges of unemployment and social security, and economic development. (e.g. APRILab, SubUrbanLab, URB@EXP).

5. ULL and knowledge sharing networks

ULL often seek to expand their "ecosystem" to broader knowledge sharing networks, such as the International Sustainable Campus Network (ISCN) and the ENoLL. These networks are co-evolving with ULL, providing a platform to develop and share standardised evaluation criteria, offering services to members with benefits through guidance, increased visibility and therefore broader impact generation. For example, JPI Urban Europe provides a platform to co-create a joint research strategy and build links between relevant stakeholders. ENoLL, ISCN and JPI Urban Europe provide a vital web between and across ULL to ensure the learning is shared and utilised. However, there is a need for further investigation into how ULL utilise such networks and the extent to which lessons and insights spread through the networks (Schliwa, 2013).

The aim of JPI Urban Europe is to "create attractive, sustainable and economically viable urban areas, in which European citizens, communities and their surroundings can thrive". At the moment, very little is known about private sector involvement in the design, implementation and evaluation of ULL presented in this chapter. In particular, there is a need for further research on if and how solutions and socio-technical change will be economically viable and embedded beyond the funding period offered by JPI Urban Europe. Long-term engagement by the private sector is needed to support ULL but there are trade-offs if ULL become too dependent on private sector funding and interests. As a European network for ULL, JPI Urban Europe plays an important role in providing funding that catalyses collaborations and development of ULL, particularly with the private sector.

Since the projects and ULL financed by JPI Urban Europe are diverse in their approach and the challenges they seek to address, a standardisation of evaluation criteria such as that provided by the ISCN might not be appropriate. Continuing the approach of co-creation and experimental learning, the evaluation of a project and application of the living labs approach may be a better subject for the people involved and affected by the project in order to maintain resilience and overcome potential lock-in created by the underlying initial vision. This is a dynamic process that will evolve over time, but is important to ensure the space for co-creation remains embedded into urban governance into the future.

Another form of knowledge sharing for ULL is the appearance of projects and networks that bring together multiple ULL from across different countries. For example, the SubUrbanLab project involves six ULL with three in Alby (Sweden) and three in Peltosaari (Finland). There is sharing of experiences and joint evaluation across theses ULL. Another example is the SusLabNRW project that aims to reduce energy consumption in the home environment. It offers an international infrastructure of living labs in four countries (Sweden, Germany, The Netherlands and the UK) to test and develop innovative sustainable concepts, products and systems. Both SubUrbanLab and SusLabNRW recognise the value of knowledge sharing across ULL in different contexts.

6. Conclusion

This chapter has examined how the concept of ULL is being designed and operationalised in contemporary urban governance for sustainability. This task was addressed by reviewing the literature and exploring five projects funded by the JPI Urban Europe, which include some 20 examples of ULL. The chapter focused on the design, establishment and goals of cutting-edge ULL initiatives within their broader urban context. As suggested, ULL are emerging as a form of collective urban governance and experimentation to address a range of sustainability challenges experienced in cities and urban areas and to capture opportunities created by urbanisation.

In terms of urban governance, ULL are clearly not entirely new but neither are they neatly continuous with past approaches. While they fit into the longer term "institutionalisation of innovation" under the neoliberal logic of urban competitiveness, they also promise a more inclusive and open process of experimentation that is capable of addressing pressing urban policy agendas surrounding climate change and smart governance. As a result, ULL are being inserted into and overlaid onto existing urban governance structures, practices and networks. Distinctive features of ULL design identified from the literature include their high visibility, which potentially fosters socio-technical transformation (1); the fact that ULL territorialise urban innovation at a manageable scale (2) and that ULL represent a specific form of experimentation with formalised processes of innovation and learning (3), which are mainly delivered by knowledge partners (4). These features indicate the importance of how ULL are designed, in determining who is involved, how problems are framed, where they are located and what broader policies or decisions they are both intended and able to influence.

Departing from the ULL design features listed above and based on the investigation of five case studies, five key characteristics of ULL were identified. These include geographical embeddedness, experimentation and learning, participation and user involvement, leadership and ownership and evaluation and refinement. Through analysing the ULL used by the selected projects, four key topics are found relevant to discuss their similarities and differences: the ways in which the ULL approach is operationalised, the type of partnership in ULL and the role of knowledge partners, the types of challenges and topics addressed by different ULL and the role of sustainability, environment, smart and low carbon agendas in the ULL.

What is clear across the cases is that ULL are bringing existing constellations of urban actors together in new ways to create more collaborative and experimental ways of "doing" urban development. There is considerable evidence that ULL are being used to address a broad range of traditional urban challenges surrounding economic development and social capacity in specific places, rather than simply focusing on smartness, sustainability and decarbonisation. A key question warranting further research involves the extent to which this type of urban development extends beyond individual projects to become embedded in existing modes of governance. The answer to this question will define their longer term impact.

References

Börjeson, M. (2008). CoreLabs. AMIWork communities wiki. Retrieved from www.ami-communities.eu/wiki/CORELABS (accessed 21 January 2015).

Bulkeley, H. and Castán Broto, V. (2013). 'Government by experiment? Global cities and the governing of climate change'. *Transactions of the Institute of British Geographers, 38*, 361–375.

EC. (n.d.). Open and participative innovation (WWW Document). European Commission's Digital Agenda for Europe. Europe 2020 Initiative. Retrieved from https://ec.europa.eu/digital-single-market/open-and-participative-innovation (accessed 13 February 2015).

European Network of Living Labs (ENoLL). (n.d.). ENoLL: About us. What is a Living Lab? European Network of Living Labs. Retrieved from http://openlivinglabs.eu/aboutus (accessed 11 November 2016).

Evans, J. (2011). 'Resilience, ecology and adaptation in the experimental city'. *Transactions of the Institute of British Geographers, 36*, 223–237.

Evans, J. and Karvonen, A. (2010). 'Living laboratories for sustainability: Exploring the politics and epistemology of urban transition'. In H. Bulkeley, V. Castán Broto, M. Hodson and S. Marvin (Eds.), *Cities and low carbon transitions*. London: Routledge.

Evans, J. and Karvonen, A. (2014). ' "Give me a laboratory and I will lower your carbon footprint"! Urban laboratories and the governance of low-carbon futures'. *International Journal of Urban and Regional Research, 38*, 413–430.

Hansen, J.R., Savini, F., Wallin, S. and Mäntysalo, R. (2013). Cases description: Preliminary exploration of case selection. APRILab: Action Oriented Research on Planning, Regulation and Investment Dilemmas in a Living Lab Experience. Retrieved from file:///C:/Users/iiie-yve/Downloads/cases_descriptions-1.pdf (accessed 11 November 2016).

Hellström Reimer, M., McCormick, K., Nilsson, E. and Arsenault, N. (2012). 'Advancing sustainable urban transformation through living labs: Looking to the Öresund Region'. Paper presented at the International Conference on Sustainability Transitions, Copenhagen, August 2012.

Hillgren, P.-A. (2013). 'Participatory design for social and public innovation: Living Labs as spaces for agonistic experiments and friendly hacking'. In E. Manzini and E. Staszowski (Eds.), *Public and collaborative: Exploring the intersection of design, social innovation and public policy*. Milan: DESIS Network. Retrieved from http://nyc.pubcollab.org/files/DESIS_PandC_Book.pdf (accessed 11 November 2016)

Hodson, M. and Marvin, S. (2007). 'Understanding the Role of the National Exemplar in Constructing "Strategic Glurbanization" '. *International Journal of Urban and Regional Research, 31*, 303–325.

Hodson, M. and Marvin, S. (2009). 'Cities mediating technological transitions: Understanding visions, intermediation and consequences'. *Technology Analysis and Strategic Management, 21*, 515–534.

Juujärvi, S. and Pesso, K. (2013). 'Actor roles in an urban living lab: What can we learn from Suurpelto, Finland?' *Technology Innovation Management Review, November*, 22–27.

Kommonen, K.-H. and Botero, A. (2013). 'Are the users driving, and how open is open? Experiences from living lab and user driven innovation projects'. *Journal of Community Information*, 9(3).

König, A. and Evans, J. (2013). 'Experimenting for sustainable development? Living laboratories, social learning, and the role of the university'. In A. König (Ed.). *Regenerative sustainable development of universities and cities: The role of living laboratories*. Cheltenham, UK, and Northampton, MA: Edward Elgar Publishing.

McCormick, K. and Kiss, B. (2015). 'Learning through renovations for urban sustainability: The case of the Malmö Innovation Platform'. *Current Opinion in Environmental Sustainability*, *16*, 44–50.

Menny, M. (2016). 'Users as co-creators? An analysis of user involvement in urban living labs'. Master thesis, Lund University. Retrieved from http://lup.lub.lu.se/luur/download?func=downloadFile&recordOId=8893215&fileOId=8893217 (accessed on 11 November 2016).

Nevens, F., Frantzeskaki, N., Gorissen, L. and Loorbach, D. (2012). 'Urban transition labs: Co-creating transformative action for sustainable cities'. *Journal of Cleaner Production, 50*, 111–122.

Percy, S. (2008). 'New Agendas'. In C. Couch, C. Fraser and S. Percy (Eds.), *Urban regeneration in Europe*. Hoboken, NJ: John Wiley and Sons.

Rösch, C. and Kaltschmitt, M. (1999). 'Energy from biomass – Do non-technical barriers prevent an increased use?' *Biomass Bioenergy*, *16*, 347–356.

Schliwa, G. (2013). 'Exploring living labs through transition management – Challenges and opportunities for sustainable urban transitions'. Master's thesis, Lund University, Sweden. Retrieved from http://lup.lub.lu.se/luur/download?func=downloadFile&recordOId=4091934&fileOId=4091935 (accessed 11 November 2016).

Ståhlbröst, A. (2008). 'Forming future IT – The living lab way of user involvement'. Doctoral thesis, Luleå University of Technology. Retrieved from http://epubl.ltu.se/1402-1544/2008/62/LTU-DT-0862-SE.pdf (accessed 11 November 2016).

Trencher, G., Bai, X., Evans, J., McCormick, K. and Yarime, M. (2014). 'University partnerships for co-designing and co-producing urban sustainability'. *Global Environmental Change, 28*, 153–165.

Veeckman, C., Schuurman, D., Leminen, S. and Westerlund, M. (2013). 'Linking Living Lab characteristics and their outcomes: Towards a conceptual framework'. *Technology Innovation Management Review, December*, 6–15.

Voytenko, Y., McCormick, K., Evans, J. and Schliwa, G. (2016). 'Urban living labs for sustainability and low carbon cities in Europe: Towards a research agenda'. *Journal of Cleaner Production, Advancing Sustainable Solutions: An Interdisciplinary and Collaborative Research Agenda, 123*, 45–54.

3

PUTTING URBAN EXPERIMENTS INTO CONTEXT

Integrating urban living labs and city-regional priorities

Mike Hodson, James Evans and Gabriele Schliwa

1. Introduction

In this chapter, we examine how urban living labs (ULL) and experiments relate to established urban governance structures and priorities. We are concerned with the ways in which ULL and experiments connect to each other within a city-wide governance context, and how this context both shapes, and is shaped by, experimentation. By considering experimentation within its broader urban context, the chapter sheds light on the potential for portfolios of experiments to shape and drive urban transformation, and by extension, the durability of experimentation as a governance strategy.

To address this, we focus on experimentation with sustainable mobility in Greater Manchester in the north-west of England. By Greater Manchester, we are referring to a metropolitan area of 2.7 million people encompassing 10 local authority areas around the city of Manchester. Greater Manchester has been positioned as a sustainable transport leader in the United Kingdom, responsible for "the largest transport infrastructure investment programme outside London", but has an acknowledged lack of capacity to deliver that programme. We focus on one element of the sustainable transport programme, cycling and efforts to constitute capacity through experimentation. Cycling has, since 2011, been incorporated in to city-regional governance structures and strategies in Greater Manchester.

The intersection of significant infrastructure investment, limited capacity to deliver and new metropolitan governance arrangements has prompted substantial experimentation with cycling infrastructure and ways of organising and governing this in Greater Manchester. This has resulted in a range of experiments within the city's designated ULL, a "Smart District" focused on the Oxford Road Corridor, and a programme of infrastructure retrofit across the city. In this case, the Oxford

Road Corridor ULL provides a platform to enable a greater concentration of infrastructure experiments.

Drawing on four years of research in this area, we (1) identify a widespread acknowledgment that these multi-actor processes and contexts were seen as experimental, (2) illustrate that discretion in designing experiments was often limited by pre-design commitments to national and city-regional scales, (3) highlight that although labs and experiments were seen as part of wider city-regional strategic priorities their realisation lacked coordination, and (4) reflect on the fact that although there were many formal and informal processes of learning about these multiple experiments, it is not clear how these were coordinated or informed strategic decision making.

The chapter has four sections. In the next section, we set out the problematic of the chapter, the relationship between ULL/experiments and urban governance priorities. In Section 3, we explore this problematic through research on cycling, ULL and experiments in Greater Manchester. Section 4 sets outs key lessons from our Greater Manchester case. Section 5 summarises our argument.

2. Connecting experiments, ULL and urban priorities

It is now well established that

> the governance concept calls our attention to the fact that governing in the urban realm involves more than "city hall" as a metaphor for monocentric government in the city. "Local government" is no more (and no less) than one constituent part of "urban governance".

Its boundaries are difficult to locate but "governance usually refers to the steering of service domains or problem areas characterized by interdependence among various involved parties and organizations" (Hendriks, 2014, p. 555). The important point being:

> Rather than thinking in terms of a singular replacement of one thing by another, we should be thinking in terms of varied shifts in more or less institutionalized working arrangements, involving both new and old types of steering, both nonformal and formal rules, and both horizontal and vertical types of relationships.
>
> (Hendriks, 2014, p. 558)

What the basis of such arrangements are and what the balance is between the strategic and the operational and between social and economic elites, on the one hand, and "ordinary citizens" (and representatives) on the other is important. That is to say, whether decisions are the outcome of "integrative deliberation" (building collective decisions through dialogue – in our case, both informing experimentation and informed by the lessons of experimentation) or more elite-led, selective forms of decision-making is a critical issue in understanding relationships between urban governance structures and priorities and forms of urban experimentation (Hendriks, 2014).

The local development of cycling capacity can be understood as taking place through forms of experiment. There are many different ways of thinking about experimentation in urban settings (Bulkeley and Castan Broto, 2013; Evans, 2011; Hodson and Marvin, 2013; Karvonen and Heur van, 2013). Here, we can think of cycling initiatives as:

> experiments [that] serve to create new forms of political space within the city, as public and private authority blur, and are primarily enacted through forms of technical intervention in infrastructure networks, drawing attention to the importance of such sites in urban climate politics.
>
> (Bulkeley and Castan Broto, 2013, p. 361)

In doing this, "experiments require a reworking of the flows of power, resources and materials through which infrastructure systems are sustained" which "requires a process of 'metabolic adjustment' through which experiments may be embedded within particular circulations and reconfigure the infrastructural 'lattice' of the city" (Castan Broto and Bulkeley, 2013, p. 1936).

ULL have emerged as the preferred venues in which cities seek to stage multiple experiments related to the challenges of climate adaptation, sustainable urban development and smart technology (Schliwa and McCormick, 2016; Voytenko, McCormick, Evans and Schliwa, 2015). In relation to the low carbon agenda, numerous experiments have been established in cities in the last 20 years, but they have tended to remain rather discrete one-off successes. As a result, funding bodies, policymakers, charities, companies and communities are increasingly focusing on ways to translate discrete experiments into broader change (Evans, Karvonen and Raven, 2016). Within this context, ULL have emerged as a mode of governance that brings stakeholders together to experiment and produce solutions in real world settings. By physically delimiting regions of a city, ULL create the conditions for potential quadruple helix partnerships between government, industry, the public and academia to address specific challenges in a specific place. Generating new technologies and services in the real world context of needs and application promises relevance and legitimacy, and has led them to be increasingly positioned as key drivers for low carbon and sustainability transitions (Evans and Karvonen, 2011). Unlike urban experimentation more generally, ULL formalise, concentrate and territorialise processes of experimentation.

3. Experimenting with cycling in Greater Manchester

The transport system in Greater Manchester is constituted by multiple transport networks that are supported by complicated governance arrangements that have evolved over a 40-year period. These arrangements have been shaped by the relationship between national government and the Greater Manchester tier of governing.

The governance of urban transport systems in the United Kingdom is a complex mix of national and urban interests and public and private interests. This means that the ability to shape these systems and the interests that do so are numerous and, depending on the particular system, there is a greater and lesser ability of interests based in Greater Manchester to shape change. In Greater Manchester, the transport system is constituted by multiple transport networks; road, rail, bus, light rail and cycling. This includes over 9,000 km of roads. Responsibility for these roads is shared between the national agency, Highways England, and local authorities.

The national rail system in the United Kingdom was privatised in 1994/1995. Total government funding of the rail industry at the time of privatisation in 1994/1995 was £1,967 million. This had increased to £5287million by 2013/2014 (Office of Rail Regulation, ORR).[1] The railway network is subject to overall governance by the national Department for Transport (DfT) and Transport Scotland (TS). They set out a five-year framework and budget with targets. The railway is operated by a combination of private, franchised train operating companies (TOCs), operating different routes, and which lease trains from rolling stock companies (ROSCOs); a public body, Network Rail, manages the tracks and railway infrastructure, including bridges, tunnels and level crossings; the ORR is the regulator for the railways (How the rail industry works – An overview, 2013). Within this, the rail network in Greater Manchester accounts for more than 25 million journeys annually, both within the city-region and to destinations north, south, east and west, a figure that has been on an upward trend. The vast majority of public expenditure on rail in Greater Manchester is passed from national government and on to private operators. This reflects the limits to public city-regional involvement in the operation of the rail network. The Metrolink light rail system, which is in the midst of a significant expansion of the network, carries around 34 million passengers a year. This system is operated and maintained by a contractor on behalf of the Greater Manchester Combined Authority (GMCA) (see GMCA/TfGM, 2011, 2017; Warriner, 2014).

The landscape of bus operators across the United Kingdom is complicated with local bus networks being operated by different configurations of private operators. The organisation of the bus industry has been shaped by the deregulation and privatisation of a system that was, until 1986, under state control via around 50 local authority operators and seven Passenger Transport Executive (PTE) owned bus companies. The significance of this is that general principles of competition law were enshrined in the operation of the bus system (White, 2010) and the control of local authorities over bus provision was greatly weakened. Tension between a competition-based market model and a gradual development of tools for local transport authorities to build coordination has been apparent since the 1990s. Within this context, the bus network in Greater Manchester is organised around radial routes into Manchester city centre and also around other town centres in the city-region. From a peak of around 400 million passengers a year in 1974, the bus network in Greater Manchester carries around 209 million passengers annually.

Since the development of a national cycling strategy in 1996, there has been a renewed policy emphasis that has aimed to increase cycling. The UK government's 10-year transport plan, in 2000, aimed to triple cycling trips in a decade. This failed, where the number of cycling trips showed no significant rise (Aldred and Jungnickel, 2014). There was not really cycling policy in the United Kingdom to speak of prior to the 1990s and when it did subsequently emerge it did so in the context of a hollowed out neo-liberal state. The consequence of this was that "cycling became embedded in public policy only after policy making had been variously outsourced to private, quasi private, and voluntary organisations" (Aldred, 2012, p.95). Additionally, there was a developing role by the new millennium for local authorities who were expected to produce a cycling strategy, which, in the logic of state hollowing out, were often outsourced to consultants (Aldred, 2012). Public investment in cycling infrastructure and training is difficult to ascertain in the United Kingdom as policy on cycling has been stop/start over the last three decades and investment in cycling and cycling infrastructure is often incorporated in funding schemes that have a wider focus than cycling. That said, it has been claimed that in the Netherlands spending per capita on cycling is between £10 and £20, while in England this figure is less than £1.[2]

At the Greater Manchester level, a new strategic governing authority –GMCA – was established in April 2011 with control over transport delegated to a new implementation body, Transport for Greater Manchester (TfGM). A statutory (usually five years) transport plan sets out the strategic direction for transport in Greater Manchester. Given the fragmented state of transport governance and funding, set out above, the financial and operational capacity for delivering on that has to be stitched together through national government grants, a levy of Greater Manchester local authorities and various other grants and schemes alongside local networks of experimentation with transport infrastructure innovation. By 2013/2014, capital spending on transport contributed to the "biggest transport capital programme outside London" (GMCA/TfGMC, 2013), but was characterised by an "overall lack of municipal and regional institutional capacity in urban and climate planning" (MacKillop, 2012, p. 249).

TfGM assumed strategic responsibilities for active travel, including cycling, in 2011. Beginning with limited internal capacity, national government funding was secured from the Local Sustainable Transport Fund in 2011 (amounting to around £23 million for cycling). A separate Cycle City Ambition Grant from national government in 2013 added a further £20 million. Additionally, there are other smaller pots of funding for cycling in Greater Manchester from, for example, national government and private rail companies for cycling training and cycling infrastructure at rail stations. This funding contributes to infrastructure projects that include creating a network of cycle routes particularly connecting employment, leisure and education sites and improved and secure cycle parking facilities at rail and transport interchanges (GMCA/TFGM, 2013).

At a Greater Manchester level, particularly given that strategic responsibility only came in 2011, there is recognition of the need for agreed standards and a unified approach to cycling infrastructure. Aside from infrastructure there is work on cycle training and information campaigns. This all contributes to the aim that between 2013 and 2025, the current proportion of trips by bicycle in Greater Manchester of about 2.2 per cent should "double and double again", producing a 10 per cent mode share by 2025 (GMCA/TfGM, 2014). What seems to be an ambitious goal is however considerably lower than what advocates suggest is required. Local cycling campaign groups, for example ask for 20 per cent mode share in their cycling manifesto and the Propensity to Cycle Tool (PTC) indicates an actual potential for 22 per cent mode share if various areas of Manchester would "Go Dutch" (Love Your Bike, 2015; Lovelace et al., 2015).

Some of Greater Manchester's 10-city councils host so-called "cycle forums" to "discuss cycling-related issues and promote opportunities for cycling" (Manchester City Council, 2016). These forums take place on a regular basis in Manchester, Salford, Stockport and Trafford and are attended by local cycling groups and councillors, members of the public as well as TfGM representatives. According to Greater Manchester's recently published Bike Life Report – a study into the current state of cycling inspired by the Copenhagen Bicycle Account – these four councils also show higher than average cycling activity compared to Greater Manchester and its other boroughs (Sustrans, 2015). The key issue, beyond discussion, is how cycling initiatives are made concrete and tangible. The issue for TfGM in relation to cycling was how to begin to build capacity to act in an area they had not historically had responsibility for. This is where TfGM found itself in 2011.

4. The Oxford Road corridor and socio-technical experiments in Greater Manchester

The Oxford Road corridor is Manchester's ULL and constitutes the highest concentration of socio-technical experiments in Greater Manchester. Corridor Manchester was formed in 2007 to generate growth and investment within the area, and was the first partnership of its kind in the United Kingdom. The corridor district (Figure 3.1) is central to Manchester's knowledge economy and home to numerous knowledge-intensive enterprises and organisations, including Manchester City Council, The University of Manchester, Manchester Metropolitan University, Central Manchester University Hospitals NHS Foundation Trust, Bruntwood (property developer), Manchester Science Partnerships, ARUP and The Royal Northern College of Music. The corridor itself is a 243 hectare non-residential area running south from St Peter's Square to Whitworth Park along Oxford Road with 70,000 students and 60,000 workers.

The Partnership's core objective is to maximise the economic potential of the area by harnessing the investment currently being made by key institutions (Universities, the Health Trust and the private sector); by stimulating future improvement and growth at key locations within the area; and by capturing economic benefit from this investment for disadvantaged local residents in the wards surrounding the area and in the city as a whole.

The Oxford Road corridor is slated to become a "physical global exemplar of knowledge based growth" (Corridor Manchester, 2015) through strategic capital investments based on five integrated themes: transport; environment and infrastructure; research and innovation; employment, business and skills; and sense of place (Corridor Manchester, 2015). Over the coming years, the corridor will receive significant upgrades to the transportation and communication networks, high-tech business activities, cultural amenities and effectively double the number of workers in this part of the city. These upgrades are intended to maximise the economic potential of the city's knowledge base, adding value to the £1.5 billion of capital investment that is committed or planned on the corridor by the main three partners over a five-year period (Corridor Manchester, 2015).

The objective of Corridor Manchester is, by 2025, for the district to be become "Manchester"'s cosmopolitan hub and world-class innovation district, where talented people from the city and across the world learn, create, work, socialise, live and do business; contributing to the economic and social dynamism of one of Europe's leading cities (Corridor Manchester, 2015, p. 4). The motivation for establishing the corridor was to leverage £1.5 billion of strategic capital investments planned by the partners to maximise the benefits to transport, environment and infrastructure, research and innovation, employment, business and skills and sense of place. Smart city initiatives that increase the social and environmental sustainability of the corridor are expected to play an important role in achieving this objective. As the most recent strategy states, "We will transform the physical environment and infrastructure of the corridor. We want to create a living and breathing laboratory . . . which will help inform future policy in the fight against climate change" (Corridor Manchester, 2015, p. 12). For example, the transformation of a section of Oxford Road to limit private traffic is part of the Bus Priority Package and has created an opportunity to promote a modal shift to public transport, cycling and walking within the corridor. The corridor has also established a global reputation by attracting high profile H2020 and Innovate UK demonstration projects.

Cycling is central to the corridor's objectives to reduce emissions of air quality pollutants, traffic congestion and carbon emissions, as well as improving the general quality of the environment for those using the corridor. While the third Local Transport Plan published in 2011 established a direction of travel for the city towards low carbon transport (GMCA/TfGM, 2011), previous efforts had been stymied, most notoriously with the rejection of the congestion charging scheme and the attached £2.8 billion of funding for public transport by popular vote in 2008. Given its privileged ability to host experimentation, the Oxford Road corridor became an obvious site to trial new types of cycling infrastructure in the city. Using the Bus Priority scheme to implement Dutch-style cycle lanes has been pushed by local cycling campaigners. However, additional road capacity has been built to accommodate the diverted traffic on a parallel street (NHS Foundation

Trust, 2014), showing that the commitment to promote cycling is not synonymous with a reduction in motorised traffic. Still, Manchester is seeking to initiate growth in cycling from a low modal share (2 per cent) and is targeting more participation from inexperienced cyclists and so plans to build more physically segregated cycle lanes (GMCA/TfGM, 2013). For a city such as Manchester to see more cyclists on its streets it has to understand the needs of cyclists, experiment with solutions and learn what works. Combined with the fragmented nature of funding described in the section above, this led the city to adopt an explicitly experimental approach to cycling. The rest of this section describes some of these experiments, starting from those that have been staged on the Oxford Road corridor as explicit experiments

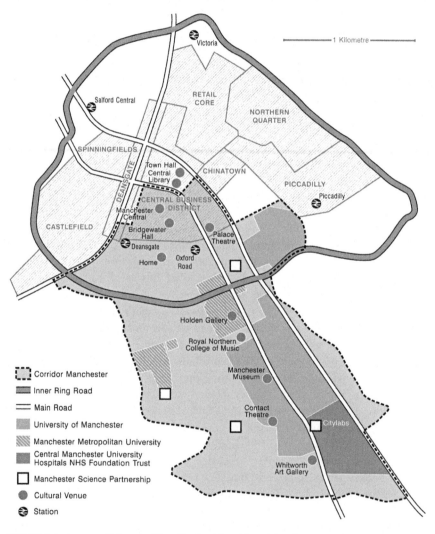

FIGURE 3.1 A map of the corridor district (Corridor, Manchester, 2015)

that are intended to be learnt from, and broadening out to consider what could be termed more "accidentally" experimental activities across Greater Manchester.

The most high-profile experiment involves the installation of Dutch style bus stops as part of the bus priority scheme on Oxford Road, whereby cycle lanes are routed behind the bus stop to avoid buses cutting across cyclists when they have to stop (Bowden, 2015). While commonplace in many other European cities, this has been considered highly experimental by Manchester transport planners. The single version of this bus stop that was installed in late 2015 is being surveyed and monitored in order to improve the next iterations and future bus stop designs. There is a considerable back story of involvement by LoveYourBike, the local Friends of the Earth cycling campaign group, who petitioned TfGM heavily during the design phases of the new Oxford Road Bus Priority scheme layout for Dutch style cycle lanes and bus stops. Rather than exerting political pressure, their activities were designed around engagement with the materiality of the urban environment, which included walking the route with planners to show them that there was enough space to implement the solution, and bringing Dutch planners to Manchester to show how it could be implemented in the city. The success was celebrated with a Friends of the Earth giant bike float at the Manchester Day Parade in 2015 entitled "Go global, go Dutch" Bicycle Block (Manchester Friends of the Earth, 2014). At the end of 2016, the Dutch style bus stops were almost complete, in addition to physically segregated cycle lanes. With monitoring in place, it will be interesting to see how they are officially evaluated over the coming years, and whether the privilege of being staged in this ULL lends this experiment extra traction.

Other parts of Greater Manchester have also been used to stage infrastructure experiments. In Salford the "armadillo" cycle lane segregation method, which involves bolting small hard plastic lumps (about 9-cm high and a similar size and shape as armadillos) to the road, has been trialled. It is a cheaper solution than segregating cycle lanes with raised kerbs, but user surveys indicated that it was not the preferred option for cyclists and this solution has subsequently been abandoned. While reported as a "failure" by local activists (MadCycleLanesOfManchester, 2014), it represents a successful trial in that it produced learning that informed decision-making. Infrastructure experiments have also been staged by non-state actors. Sustrans, a national sustainable transport NGO, facilitated an experimental street design in Stockport inspired by tactical urbanism, which involved painting zigzag stripes and flowers on the road to make cars drive more slowly. Since Sustrans was not officially allowed to change road markings, adding flowers turned it into street art which made it acceptable to the local authority (Duran, 2015).

One of TfGM's flagship cycling experiments entails the establishment of "cycling hubs" in Manchester city centre and in Salford Quays to provide secure bike parking and shower facilities to encourage office workers to commute by bike. Run along similar lines to commercial gyms, different membership options exist. The space for the first hub in City Tower, Piccadilly Gardens, was provided by one of the city's largest property owners, Bruntwood, and a new hub has opened in Media City, Salford Quays. Opinion is split over whether they represent the future of commuter cycling or misguided corporate green-washing, with rumours suggesting that the hubs are half empty for a majority of the time (Wilson, 2015).

A number of research projects have been spun off from the £20 million Velocity cycling investment programme that have also focused on the Oxford Road corridor ULL. The Manchester Cycling Lab, funded by the Economic and Social Research Council (ESRC) and run out of The University of Manchester, conducted a 360° analysis of the key stakeholder knowledge needs relating to the Velocity objectives in the Oxford Road corridor. The project linked seven research students at the University with key stakeholders in a strategically coordinated way to maximise the collective impact of research capacity (Manchester Cycling Lab, 2015; Sheldrick, Evans and Schliwa, 2016). The project team also sought to test out the potential to engage users in cycle infrastructure planning through a range of digital engagement techniques (Schliwa, Bell, Evans and Mason, 2015). This work provided the basis for a cycling logistics project in the Oxford Road corridor, using electric cargo bikes, which forms part of a £20 million European Commission Horizon 2020 Smart Cities and Communities project (Schliwa, Armitage, Aziz, Evans and Rhoades, 2015). It also forms a thematic strand of pilot sustainable transport experiments in the city's recently successful £10 million Internet of Things (IoT) Innovate UK City Verve project. These projects indicate the increasingly close involvement of academic partners in ULL, and reflect the growing emphasis on applied project research funding and, on the face of it, an appetite among urban stakeholders for specific knowledge and evidence to support decisions that research partners are seen as being able to provide.

5. Cycling experiments and strategic priorities in Greater Manchester

From the examples set out above and drawing on interview material, we highlight a set of issues that are raised in the relationship between ULL/experiments in Greater Manchester and formal urban governance structures. These issues are indicative rather than exhaustive and are intended to contribute to our understanding of experiments, the parametres on action that experiments set, the relationships that experiments have to each other and how they connect to wider city-regional priorities in Greater Manchester. The four issues are:

i.   **There was recognition that urban cycling initiatives were experimental.**

There was recognition from a wide range of interests, in both initiatives and formal governance settings, that different initiatives were seen as experimental. A cycling activist suggested: "A lot of it [Greater Manchester cycling projects] is experimental at the moment". In the specific case of a city-regional policymaker reflecting on a cycling hub development:

> There's an element of experimentation here, if a very basic facility wasn't providing the answer for what people wanted, would a quality facility

provide that? I think we probably answer, "Yes". In both cases, the proof of the pudding would be the success of the hubs.

There was a clear geographical element to this definition – things that had not been tried in Manchester before were experimental regardless of how commonplace they may have been elsewhere. An experimental approach was also necessitated by the need to constitute the capacity to deliver cycling infrastructure projects. Initially when they started most of the cycle lane designers, for example were primarily highway engineers and bus planners.

ii. **Constrained discretion.**

There was also a sense from a number of interests that the ability to exercise discretion in experimentation was limited given commitments that had been made to funders, most notably at the national and European levels, prior to development of material infrastructures: "Part of the problems we had with them [the transport authority] was they rigidly enforced it afterwards when others [places] that got money seemed to be quite happy to swerve a little bit and change things". This extends to the nature of the partnership itself. The Cycle Hub scheme has been criticised for being too corporate in its approach.

iii. **A lack of coordination of experiments?**

The reliance on multiple funding schemes produced a culture whereby although there was strategic orientation, "In terms of delivery we always see pieces".

What this points to is a lack of coordination between different experiments. What there isn't, there's no co-ordination, and the part of the problem with this stuff, particularly cycling, it crosses about six different bits, even within one council . . . there's highways, there's planning, there's leisure, there's public health, there's dah, dah, dah and there's no one person. Either at a political level or an officer level, there's no one person who oversees this stuff.

The consequence is a lack of join-up between initiatives. Though there is to some extent a systemic view of the development of cycle lanes and routes, as set out in Greater Manchester's cycling strategy, there remain issues about how infrastructure developments are designed and implemented, used but also how they are used and whether they connect to other developments, such as the rapidly expanding Metrolink tram network.

Experiences on the corridor ULL reinforce this view, with an explicit focus on project-based innovation sitting uneasily alongside grand aspirations to systemically transform the physical environment. The fragmentary character of urban experiments has been described most fully in relation to Masdar City, a well-known experimental eco-city project under development in the United Arab Emirates (Cugurullo, 2016). Masdar City is understood as an urban experiment composed of multiple sub-experiments such as smart grid installations, large-scale pedestrian spaces, fusions of vernacular and modern architectural styles, and automated transport systems. The number of actors involved, divergent understandings of sustainability and fragmented projects ultimately prevents Masdar City from achieving its vision of the sustainable city. Greater Manchester reveals similar tensions between urban experimentation and sustainability in an existing urban context.

iv. **Experiments and existing governance context.**

A further issue relates to how knowledge of and learning from ULL and about experimentation is built at Greater Manchester level. There is a cycling strategy (and various other transport strategies) and there are also participatory processes for producing such strategies and refreshing them, such as the cycle forums as well as public consultations. But in terms of a wider understanding of how successful experiments are and how they contribute to urban governance strategies and priorities this is less clear:

> it's just hard to grasp what's been delivered, how successful it is. I mean I wouldn't really know how successful [it] has been for Manchester. There's a difference between saying you've engaged 1,000 or 10,000 businesses compared to what have you delivered and what's the outcome.

There are very many formal and informal processes of measuring and learning (e.g. follow-up questionnaires on projects; different forms of official data) about multiple experiments but it is not clear how these multiple processes are coordinated and feed back into strategic decision-making: "This opens up a huge can of worms. The honest answer is they can't". There are also more qualitative processes of learning which are even more difficult to formalise. That there is limited progress in this regard can best be seen through anecdotal evidence concerning corporate memory. "It's something that we face all the time, even before the austerity started. People would leave or get a different job or be promoted or demoted or fired or what have you, there's little corporate memory". This was the case in various parts of the governance configuration where, for example transport authority staff left, retired or moved internally, the Velocity project manager within Manchester City Council changed job, and the Manchester Cycle Forum coordinator left. While not uncommon in public bodies, this specifically hampers learning from experiments in the absence of a new framework for strategic planning.

Reflecting on the role of the Oxford Road corridor as an ULL intended to facilitate such learning, it is clear that while it has focused experimental efforts, the degree to which it can coordinate their results in such a way to make the whole greater than the sum of its parts and drive broader change remains limited. Beyond the challenge of coordinating the different organisations and initiatives that collect data on experiments, the task of feeding any combined learning into policy and decision-making often founders on the disjuncture between experimental and formal governing approaches. The popularity of the kinds of Urban Data Platforms and open data hubs found in the current smart city discourse is that it promises a technical fix for this inherently political challenge. In reality, what we are seeing is a structural disjuncture between traditional top-down urban planning and bottom-up partnerships that remake the city one project at a time. In terms of transport, this may reflect the corridor's composition, which does not include the transport body, TfGM, in the partnership. In turn, this is a consequence of the corridor's origins as a territorial pact between adjacent, large estate owners. The answer to whether the outcome would be any different for experiments focused on energy is "probably not". In this sense, the corridor is a space waiting to be filled, albeit inhabited by organisations who have stated a willingness to host experiments. A key question here becomes how to tie sector specific organisations effectively into ULL.

6. Conclusion

In this chapter, we have addressed how ULL and experiments relate to established urban governance structures and priorities. We explored this issue through the case of cycling in Greater Manchester, which came to be a focus for experimental approaches in the face of fragmented financial and political support. The formal governing context in Greater Manchester has been subject to on-going experimentation over a period of four decades (Hodson and Marvin, 2013). The most recent phase of experimentation has resulted in cycling occupying a strategic role and the development of formal governing structures, priorities and plans in relation to cycling. Efforts to constitute cycling capacity in Greater Manchester were illustrated through a number of ULL and experiments. The chapter set out a series of issues and challenges in relationships between ULL, experiments and formal governing contexts. The current round of urban experimentation differs from previous incarnations, representing a specific kind of governance "fix" for a broadly neoliberal system that is struggling to move towards more sustainable forms of urban development. As a distinct mode of urban governance, the challenge for experimentation is how to bring people and infrastructures together in ways that are capable of realising significantly different and more sustainable urban futures in socially just and democratic ways (Evans and Karvonen, 2014). A key part of addressing this challenge involves developing clearer mechanisms through which ULL and experiments can influence formal governing structures and shape strategic urban agendas.

Notes

1 Government support to the rail industry – Table 1.6, 1985–1986 to 2013–2014 Great Britain (£ million) at https://dataportal.orr.gov.uk/displayreport/report/html/0913a 84d-b740-4111-b6f8-bf6470e2d7b7 (accessed 15 January 2015).

2 www.cyclingweekly.co.uk/news/how-britain-has-failed-cycling-46388 (accessed 10 February 2015).

References

Aldred, R. (2012). 'Governing transport from welfare state to hollow state: The case of cycling in the UK'. *Transport Policy, 23*, 95–102.

Aldred, R. and Jungnickel, K. (2014). 'Why culture matters for transport policy: The case of cycling in the UK'. *Journal of Transport Geography, 34*, 78–87.

Bowden, A. (2015). First bus stop bypass for cyclists installed on Manchester's Oxford Road. *Manchester Evening News*, September 25. Retrieved from http://road.cc/content/news/165773-first-bus-stop-bypass-cyclists-installed-manchester-s-oxford-road (accessed 28 January 2016).

Bulkeley, H. and Castan Broto, V. (2013). 'Government by experiment? Global cities and the governing of climate change'. *Transactions of the Institute of British Geographers, 38*, 361–375.

Castan Broto, V. and Bulkeley, H. (2013). 'Maintaining climate change experiments: Urban political ecology and the everyday reconfiguration of urban infrastructure'. *International Journal of Urban and Regional Research*vol, *37*(6), 1934–1948.

Corridor Manchester. (2015). *A decade of opportunity: Strategic vision to 2025*. Retrieved from www.corridormanchester.com/_filestore/corridormanchester/mmu1954-corridor-manchester-strategic-vision-aw-40pp-web-pdf/original/MMU1954%20Corridor%20Manchester%20Strategic%20Vision%20AW%2040pp%20WEB.pdf (accessed 29 January 2016).

Cugurullo, F. (2016). 'Frankenstein cities: (De)composed urbanism and experimental eco-cities'. In J. Evans, A. Karvonen and R. Raven (Eds.), *The experimental city: New modes and prospects of urban transformation*. London: Routledge.

Duran, J. (2015). Creativity can change our streets. On the Platform 30 July 2015. Retrieved from http://gmlch.ontheplatform.org.uk/article/creativity-can-change-our-streets (accessed 28 January 2016).

Evans, J. (2011). 'Resilience, ecology and adaptation in the experimental city'. *Transactions of the Institute of British Geographers, 36*, 223–237.

Evans, J. and Karvonen, A. (2011). 'Living laboratories for sustainability: Exploring the politics and epistemology of urban adaptation'. In H. Bulkeley, V. Castán Broto, M. Hodson and S. Marvin (Eds.), *Cities and low carbon transitions* (pp. 126–141). London: Routledge.

Evans, J., Karvonen, A. and Raven, R. (2016). 'Introduction'. In J. Evans, A. Karvonen and R. Raven (Eds.), *The experimental city: New modes and prospects of urban transformation*. London: Routledge.

GMCA/TfGM. (2011). *Greater Manchester's third Local Transport Plan 2011/12–2015/16*.

GMCA/TfGM. (2013). *Vélocity 2025*. Retrieved from http://cycling.tfgm.com/Pages/Velocity.aspx (accessed 29 January 2016).

GMCA/TfGM. (2014). Greater Manchester cycling strategy.

GMCA/TfGM. (2017). Greater Manchester Transport Strategy Evidence Base. Retrieved from https://assets.contentful.com/nv7y93idf4jq/3OOAkf1PSgQGUqiseGcOoI/09e308f5cb7e0013674e79ee7a74fa1c/04._GM_2040_TS_Evidence_base_-_Published_Feb_2017.pdf

Hendriks, F. (2014). 'Understanding good urban governance: Essentials, shifts, and values'. *Urban Affairs Review, 50*(4), 553–576.

Hodson, M. and Marvin, S. (2013). *Low carbon nation?* London: Earthscan.

How the Rail Industry Works – An Overview. (2013). Retrieved from www.transportfocus.org.uk/research/publications/how-the-rail-industry-works-an-overview

Karvonen, A. and Heur van, B. (2013). 'Urban laboratories: Experiments in reworking cities'. *International Journal of Urban and Regional Research, 38*(2), 379–392.

Lovelace, R., Goodman, A., Aldred, A., Berkoff, N., Abbas, A. and Woodcock, J. (2015). 'The propensity to cycle tool: An open source online system for sustainable transport planning'. *ArXiv:1509.04425 [Cs]*, September. Retrieved from http://pct.bike (accessed 29 January 2016).

Love Your Bike. (2015). Getting moving: A cycling manifesto for Greater Manchester. Retrieved from www.manchesterfoe.org.uk/loveyourbike/cyclingmanifesto (accessed 29 January 2016).

MacKillop, F. (2012). 'Climatic city: Two centuries of urban planning and climate science in Manchester (UK) and its region'. *Cities, 29*, 244–251.

MadCycleLanesOfManchester. (2014). Salford's armadillo consultation – 5th May. Retrieved from http://madcyclelanesofmanchester.blogspot.co.uk/2014/04/salfords-armadillo-consultation-5th-may.html (accessed 28 January 2016).

Manchester City Council. (2016). Manchester cycle forum. Retrieved from www.manchester.gov.uk/info/200102/cycling_and_walking/6851/manchester_cycle_forum (accessed 29 January 2016).

Manchester Cycling Lab. (2015). Manchester cycling lab research. Retrieved from http://universitylivinglab.org/manchestercyclinglab/research (accessed 29 January 2016).

Manchester Friends of the Earth. (2014). Manchester day parade (Sunday 22 June) – Going global – Going Dutch. Retrieved from www.manchesterfoe.org.uk/manchester-day-parade-sunday-22nd-june-going-global-going-dutch (accessed 28 January 2016).

NHS Foundation Trust. (2014). *Extended 'Dutch-style' cycle lanes part of new plans for bus priority on Oxford Road*. Latest news from 20 November 2014. Retrieved from www.cmft.nhs.uk/media-centre/latest-news/extended-%E2%80%9Cdutch-style%E2%80%9D-cycle-lanes-part-of-new-plans-for-bus-priority-on-oxford-road (accessed 29 January 2016).

Schliwa, G., Armitage, R., Aziz, S., Evans, J. and Rhoades, J. (2015). 'Ustainable city logistics – Making cargo cycles viable for urban freight transport'. *Research in Transportation Business and Management, 2015*, 50–57.

Schliwa, G., Bell, B., Evans, J. and Mason, C. (2015). 'Mobile phone apps: A rich source of data for cycle infrastructure planning?' Local Transport Today (LLT) Issue 664, *Viewpoint*, p. 18.

Schliwa, G. and McCormick, K. (2016). 'Living Labs – Users, citizens and transitions'. In Evans, J., Karvonen, A. and Raven, R. (Eds.), *The experimental city: New modes and prospects of urban transformation* (pp. 163–178). London: Routledge.

Sheldrick, A, Evans, J. and Schliwa, G. (2016). 'Policy learning and sustainable urban transitions: Mobilising Berlin's cycling renaissance'. *Urban Studies*, 1–24.

Sustrans. (2015). Bike life Greater Manchester. Retrieved from www.sustrans.org.uk/sites/default/files/bike_life_greater_manchester_2015.pdf

Voytenko, Y., McCormick, K., Evans, J. and Schliwa, G. (2015). 'Urban living labs for sustainability and low carbon cities in Europe: Towards a research agenda'. *Journal of Cleaner Production*. doi:10.1016/j.jclepro.2015.08.053

Warriner, S. (2014). 'Transport budget briefing, Town Hall, Manchester'. Director of Finance and Corporate Services, January 21, TfGM.

White, P. (2010). 'The conflict between competition policy and the wider role of the local bus industry in Britain'. *Research in Transportation Economics, 29*, 152–158.

Wilson, P. (2015). 'The development of cycle commuting centres from a social practice perspective'. Paper presentation at Cycling and Society Symposium, Manchester, September 14, 2015. Retrieved from www.cyclingandsociety.org/wp-content/uploads/2015/10/CSS2015-Paul-Wilson-Presentation.pdf (accessed 29 January 2016).

4

URBAN LIVING LABS FOR THE SMART GRID

Experimentation, governmentality and urban energy transitions

Anthony M. Levenda

1. Introduction

A variety of urban actors are increasingly experimenting with smart grid technologies to simultaneously address climate change, environmental, economic and sustainability concerns. Reconfiguring urban electricity systems to allow for more renewable energy integration, increased reliability and new forms of consumer engagement have been seen as a vital ingredient for a smart, and low-carbon, energy transition. In attempts to govern energy transitions, sociotechnical interventions are being tested out in experimental venues where new knowledge can be gained on how cities are best to react to pertinent energy sustainability issues. On the one hand, experimentation offers opportunities for radical alternatives and innovations, and on the other, it may serve only to further support existing socio-technical regimes.

This chapter engages with the concept of urban living labs (ULL) to investigate how experimentation impacts the city and its citizens. ULL are seen as normatively ambiguous modes of technological implementation in the "real-world", forms of experimentation that facilitate learning and experiential knowledge production, and social arrangements that interlink various actors from research and development to government and industry (Karvonen and van Heur, 2014; Evans and König, 2013; Voytenko, McCormick, Evans and Schliwa, 2016). The burgeoning literature and emergent policy attention to ULL suggest that they can serve not only as one-off projects for demonstration but sustained socio-technical changes that help cities achieve goals for sustainability and low-carbon transition more broadly. At the same time, however, ULL are often only ephemeral, mapping particular pathways and materialising visions, but not generative of systemic change.

ULL have been implemented to address a variety of challenges, especially relating to climate adaptation, urban sustainability and more recently, "smart" technology. At the intersection of these challenges is the urban smart grid, which promises to allow for decarbonisation of energy supply, increased grid reliability, and electrification of transport all while saving consumers money. The growth of smart city and smart grid projects has made integration of digital information and communication technologies (ICTs) a prominent part of the urban electricity system (Marvin, Luque-Ayala and McFarlane, 2015; McLean, Bulkeley and Crang, 2015).

The smart grid integrates ICTs with the electricity system to provide data on the use, distribution, supply and demand of electricity on the electrical grid. Equipped with sensors, monitors and internet-enabled digital communication devices, data circulates alongside electrons to provide information on outages, time of use, variations in loads, and energy consumption in fine-grained detail. But the smart grid is much more than an array of networked technologies. As Luque (2014, p. 160, emphasis added) explains:

> the smart grid is an assemblage of networks, technologies, and users interacting through telecommunications platforms. It is a *socio-technical intervention* that relies on utility networks, technological equipment and digital software as well as knowledge networks and an emerging set of user practices.

As a socio-technical intervention, smart grids offer a jumping off point for new relations of production and consumption, but also further investment and public buy-in for larger smart city projects including implementation of smart, networked technologies for transportation, environmental monitoring, water, waste, government operations, etc.

In Austin, Texas, an urban smart grid experiment known as the Pecan Street Project (PSP, named after the non-profit organisation, Pecan Street, which organised and implemented the project) was developed to demonstrate and learn from various socio-technical interventions including high concentrations of electric vehicles and distributed solar, smart home technologies and advanced smart grid infrastructure. Fuelled by local political leaders, a private–public partnership, and a national smart grid demonstration program (funding stimulus), the PSP was said to have created "the most innovative neighbourhood" and the "world's largest database on customer energy use" (Frangoul, 2015). By engaging environmentally concerned and/or technologically savvy residents of the Mueller neighbourhood (where the PSP is located) with incentives to adopt electric vehicles, solar panels, home energy management systems (HEMSs) and numerous other smart technologies, Pecan Street was able to get considerable participation and access to collect fine-grained data on energy usage of consumers. This data resource and development project has garnered attention from numerous city officials, technology companies and researchers interested in the smart energy transition and all of its possible benefits, including "jolting up the green economy" (Frangoul, 2015).

Austin's urban smart grid experiment discussed in this chapter configures governance *through* experimentation (Bulkeley and Castán Broto, 2013b; Castán Broto and Bulkeley, 2013a). However, experimentation is not solely about infrastructural change or urban entrepreneurialism. Rather it implies new possibilities for relationships between producers, consumers, infrastructures, and practices. This presents a need to investigate the way in which "participants" of experiments are enrolled in experimentation and how their activities, practices and conduct are as much the subject of experimentation as is infrastructural or policy changes. Decisions that shape and become embedded in ULL, and the forms of experimentation they facilitate, influence the activities and practices of urban citizens. Therefore, we must consider the politics of ULL design, their material and discursive construction, the forms of experimentation they facilitate, and related knowledge claims that are supported by them. This chapter suggests an analytics of governmentality helps us understand the power relations embedded in ULL and how urban experimentation works to reconfigure everyday practices through smart grid technologies and infrastructures.

The chapter starts by unpacking the metaphor of the ULL to understand the epistemologies underpinning urban experimentation, with particular attention to the ways in which ULL are created materially and discursively as particular places for technological demonstration and testing, so called urban test beds. Urban test beds are more limited than ULL in that they focus exclusively on technological testing and demonstration, but often claim the same moniker of a "living lab". Then I discuss the development of a smart grid experiment in Austin highlighting how the project facilitated demonstration and test-bedding approaches with aims for attracting technology companies to the city, and embedding new logics and rationalities for organising energy practices and conduct.

2. Laboratorising the city: A smart grid experiment in Austin, Texas

The City of Austin has been recognised as a leader in sustainability. Austin is known for an environmentalist ethos that emerged most prominently in the actions to preserve Barton Springs and the Edwards Aquifer during the 1970s and 1980s, an issue that aligned various actors on two sides of a long-term environmental and political struggle (Moore, 2006). Since the early 1990s, the city has worked to tackle energy and climate issues through green building programs, energy efficiency strategies, and implementation of clean energy production. With the growth in Austin's economy – focused on creative and tech industries – the city is positioned as a leader for innovation in clean tech and clean energy solutions. The large research university – the University of Texas (UT-Austin) – has also positioned Austin as an important hub for knowledge-based industries, often leveraging university partnerships for research and development. Austin's organisational structure of energy provision also helps to foster an engaged citizenry: Austin has one of the largest municipal utilities in the nation, Austin Energy, responsible to Austin's City Council. These characteristics of Texas's most liberal and progressive city created a favourable environment for the smart grid demonstration project implemented by Pecan Street, a non-profit research and development organisation.

The smart grid project, originally known as the PSP, aimed to research and learn from the implementation of various smart grid and smart home technologies – including solar panels, electric vehicles, various HEMSs, smart metres, control and visualisation technologies and energy storage technologies – implemented in an urban neighbourhood. The project started at the Mueller Development – a private–public redevelopment project that commenced in 2004 on a nearly 700-acre defunct airport base just three miles northeast of downtown Austin and the University of Texas. The redevelopment project was a source of political contestation during development and more recently during disputes over affordability and density (Clark-Madison, 2002; King, 2015; Reeves, 2008). As the LEED (Leadership in Energy and Environmental Design)-certified neighbourhood was celebrated for its sustainability, it was able to gain the political momentum needed for implementation of the development plan. The "clean slate" of the new development project also served as an ideal location for the Pecan Street ULL project.

The demands of the smart grid experiment required a physical infrastructure that was already "modernised", or in this case, built from scratch. This included the development of Austin Energy's smart grid platform (Carvallo and Cooper, 2015), new green-built homes, and Pecan Street's own information and communications or smart grid network. In addition, to test various smart grid technologies, the research participants needed to acquire various smart grid technologies – everything from solar panels and electric vehicles to smart appliances and HEMS equipped with visualisation and control technologies. The new development project in central Austin was the ideal location for this ULL also because it served as a way to recruit participants for the research Pecan Street would conduct. As one of the studies based on the demonstration project describes it: "Mueller was selected as the test-bed for this research project because of its location, the relative uniformity of new homes, and the developer's requirement to build energy efficient homes and buildings" (Rhodes et al., 2014, p. 463).

The ULL project started as a partnership with the University of Texas, the Mueller Development, the City of Austin, Austin Energy (the municipal utility), the Austin Technology Incubator and industry partners led by the non-profit umbrella organisation, Pecan Street, Inc. Now known simply as Pecan Street, the organisation has been leading research and implementation of smart grid technologies on "the consumer-side of the grid" (Pecan Street representative, Interview, October 2015). The smart grid experiment started as an "energy internet demonstration" project, spurred and supported by a Department of Energy Smart Grid Demonstration grant part of the larger American Recovery and Reinvestment Act (ARRA) of 2009. An initial chairman described the genesis of the project and explained its initial purpose and goals:

We started working and focusing on the data acquisition and manipulation side of energy, energy efficiency, renewable energy, electric vehicles and so forth. [. . .] Through that grant we were able to focus on a mass deployment of both electric vehicles and solar. We had the highest concentration of electric vehicles and solar in any neighborhood in the country. [. . .] One of the arguments we made for them was that there was plenty of capacity for the electric vehicles because people would be charging at night. Well, we actually didn't know that [laughter] and if everyone was plugging in their vehicles and charging them in the afternoon, then that adds to the problem rather than relieving it. Fortunately, the research has shown the charging is spread out pretty evenly throughout and we were able to document that and show it.

(Former Austin Energy Executive, Interview, May 2016)

As the interview quotation above suggests, the project was aimed at *proving* that smart grid technologies would allow integration of electric vehicles and distributed solar generation reliably. The data collected through the smart grid project also helped show the efficacy of interventions, with granularity providing greater insights. Just two miles northeast of downtown on of data power nderstand the impacts on privacy or equity. But these technologies are no. By focusing on energy monitoring and data collection, Pecan Street's database, analysis and sharing wiki, Dataport, the group has built a fundamental resource for energy research in smart grids. The dissemination of information via Dataport on energy consumption and production has influenced numerous lessons and possible research trajectories on energy management both by consumers and utilities (inter alia Alahakoon and Yu, 2013; Ranganathan and Nygard, 2011). This ever-growing dataset is informing research on a variety of issues positioned as solutions for numerous energy and environmental problems (such as climate change) and technical problems for utilities (such as demand side management and peak load shaving). This focus has positioned this smart grid experiment as a place for learning about how to create a viable and profitable smart energy transition.

Austin's smart grid experiment is an example of the way cities are increasingly made the node for experimentation with socio-technical interventions for sustainability and low-carbon transitions through so-called living laboratories (Evans and König, 2013), urban laboratories (Evans and Karvonen, 2013; Karvonen and van Heur, 2014) and ULL (Reimer, McCormick, Nilsson and Arsenault, 2012; Voytenko et al., 2016). The concept of the laboratory invokes the idea of a sterile, enclosed and exclusive space for knowledge production wherein scientific experiments are run separate from society (Allen, 2011; Evans and Karvonen, 2013; Gieryn, 2006; Gopakumar, 2014; Heathcott, 2005; Strebel and Jacobs, 2014). But classic work in science and technology studies (STS) – especially laboratory studies in the late 1970s – examined and unpacked the "hard core" of scientific work: its technical content and the production of knowledge (Knorr-Cetina, 1995).

The laboratory itself, as Knorr-Cetina (1995) posits, is an important theoretical notion in the social studies of science because it reveals "the power of locales in modern institutions and raise[s] questions about the status of 'the local' in modern society in general". At the same time Livingstone (2010, p. 3) suggests that the laboratory is a privileged place of knowledge production, where a concerted effort was made to create it as a "placeless place" to do scientific activities and where local contingency has no impact on those activities. A laboratory here is the same as a laboratory anywhere. This "placelessness" was also key to securing the credibility and objectivity in the production of knowledge.

The concept of placelessness poses practical difficulties for ULL, however, because they are situated in urban contexts that have histories, cultures and economies that bleed through porous boundaries presenting a complex mix of endogenous and exogenous factors contributing to observed changes. At the same time, the place of an ULL tends to influence the legitimacy and credibility of knowledge claims within and about the city (Gieryn, 2006). For example, early Chicago School urban sociology used the city variedly as a field site – an uncorrupted reality – and a laboratory – a controlled environment providing the ability for generalisations true for other cities. ULL tend to draw on the virtues of both lab and field, wherein the city becomes both the object (what) and venue (where) of study, allowing multiple modes of inquiry to make "valid" claims while "creating a discursive situation in which location, geography and situated materialities get foregrounded as ratifiers of believability" (Gieryn, 2006, p. 28). Sites where knowledge claims are made about the efficacy of urban sustainability or low-carbon transitions thus have a bearing on how urban sustainability or low-carbon transitions are defined and legitimated, and how they gain wider acceptance as models and exemplars producing "best practices" for any other place.

As a mode of governing urban socio-technical systems, such as the smart grid, ULL facilitate forms of experimentation that utilise place-based claims and claims to placelessness simultaneously. This central tension filters through the two modes of governance examined in this chapter: governing by experiment and governing energy conduct. First, governing by experiment relies on place-based claims to demonstrate the efficacy of urban experiments to provide material manifestations of possible futures, and to show how novel socio-technical configurations work in the real-world. At the same time, knowledge generated from these experiments is supposed to be useful in other places, providing a real-world "truth-spot" that attests to the legitimacy of claims. The design and siting of the ULL are important political resources for giving authority to experiments and building momentum for transitions. As Gieryn (2006, p. 28) notes, "political pronouncements have different consequences when uttered from the street corner – or from the floor of an official parliamentary space". So, while experimentation may harness the radical contingency and messiness of cities, ULL provide a structured and politically powerful space for protecting and nurturing socio-technical interventions.

Second, governing energy conduct refers to a specific form of urban governmentality that rests upon an imagined smart consumer subject. Instead of using place or placelessness to support claims, this form of governing uses the discursive space of the "laboratory" to attempt to shape actions and behaviours through appealing to particular rationalities based on neoliberal conceptions of *homo-economicus*, or economic man. Similar to Strengers (2013, p. 51) "resource man" – "a data-driven, information-hungry, technology-savvy home energy manager" – *homo-economicus* captures the economic and entrepreneurial subjectivity of energy users implied in many ULL strategies. The critical perspective of governmentality helps us locate and potentially confront the dominant neoliberal political rationality shaping urban smart grid experiments.

These forms of governing are important because they become embedded in the materiality of the urban and in the technologies and infrastructures of ULL. In particular, technological systems generate certain forms of "user scripts" – sets of "normal" and acceptable use for various technologies – which govern the way people use technologies (Oudshoorn and Pinch, 2005). These technological inscriptions are of course contestable and open to change. Technologies have "interpretive flexibility" but often are locked-in over time (Bijker, Hughes and Pinch, 1987), becoming part of larger socio-technical systems, sets of practices and political economic rationalities that make them resistant to change. As smart technologies are implemented in the context of ULL, they configure particular sets of user practices normal and acceptable, while also creating new path dependencies and reconfigurations of urban infrastructure.

3. Governing through experimentation: Technological demonstrations and test bed urbanism

Experiments are "purposive and strategic but explicitly seek to capture new forms of learning or experience . . . they are interventions to try out new ideas and methods in the context of future uncertainties serving to understand how interventions work in practice" (Castán Broto and Bulkeley, 2013, p. 93). They offer the "means through which discourses and visions concerning the future of cities are rendered practical, and governable" (Bulkeley and Castán Broto, 2013b, p. 367). The way localities respond to and govern energy challenges such as climate change and sustainability are, in part, choreographed and mediated through the experimental landscapes and infrastructures of ULL. Thus, ULL not only require consideration as spaces of knowledge production and innovation, but also as places where (the future of) cities are governed.

While experiments offer opportunities for learning and radical innovation, they also may simply reinforce existing regimes. Experiments are often driven by the motivations of powerful actors – profit, a sense of urgency to act, a desire to expand authority and express ideologies. These are all clearly visible in the ways cities are responding in experiments, and in the visions they express discursively and manifest materially (Hodson and Marvin, 2009b; While, Jonas and Gibbs, 2010). At the same time, experiments can be exclusionary, technological "fixes" for issues that are inherently social and political, reflecting particular visions of powerful actors and interests reinforcing or creating new injustices (Swyngedouw, 2011; While, Jonas and Gibbs, 2004). Experimentation offers opportunities to "open-up" the city for private investment and control of urban infrastructures. Fitting with the dominant form of neoliberal urban governance, governing through experiment aligns with the entrepreneurial role of local governments (Davidson and Gleeson, 2014; Hall and Hubbard, 1996; MacLeod, 2002). As Harvey (1989, p. 5) argued, with the turn from managerialism to entrepreneurialism in urban governance, "investment increasingly takes the form of a negotiation between international finance capital and local powers doing the best they can to maximise the attractiveness of the local site as a lure for capitalist development". Opening up the city as a test bed or a demonstration site for new smart technologies provides an opportunity to attract highly mobile capital. However, this may have "splintering" impacts in the city (Graham and Marvin, 2001) that create spaces of high value while simultaneously excluding and marginalising other spaces and communities.

In Austin, for example, the Mueller redevelopment project has struggled with the problematics of meeting and sustaining affordable housing requirements guaranteed in the original partnership and development agreement. This problematic stems from the Travis Central Appraisal District (TCAD) methods for taxation, appraising affordable housing at market rates, sometimes 100 per cent over the purchase price (King, 2015). While the rising attention to the Mueller redevelopment as a model of sustainability has captured the attention of so-called new urbanists and sustainability officers of cities around the world, it has also contributed a growing unaffordable housing market and the correlated issue of greater displacement of communities of color in east Austin (Long, 2016).

While the affordability issue of "losing properties to the market" (Representative of Mueller Development, Interview, November 2015) was battled with a non-profit organisation, the Mueller Foundation, considerable taxation increases based on market-rate appraisals were creating un-affordable conditions for many homeowners. The program's commitment to a sustained affordable homes program is laudable, but the struggle to keep housing "off the market" will likely continue as the redevelopment project continues to add new housing units into the future. And while many of the initial homeowners benefitted from the subsidies and tax breaks organised by the PSP, state government and federal government, the access to smart grid technologies (including EVs (electric vehicles), PV (photovoltaics) systems, HEM (home energy management) systems, smart appliances, etc.) may not reach the broad growing community of affordable homes buyers. The affordability issues are not only tied to the improved infrastructure of the redevelopment and the technological innovation spurred by the ULL project, but it also signals a larger trend in the political economy of Austin's urban development with the creation of "ecological enclaves" (Hodson and Marvin, 2010) for well-off sustainability-minded citizens and urban "technological zones" (Barry, 2006) for the testing and establishing network and connection standards for new ICT infrastructures.

As a sustainability-focused development with a "clean-slate" of green-built homes, and a newly modernised electrical grid, the smart grid experiment and demonstration was thought to "logically" fit at Mueller. In the new redevelopment project, infrastructures were designed to be amenable to smart grid technologies, the newly built homes were developed with a sustainability focus, and the larger vision for Mueller was to be an icon and demonstration for urban sustainability. This vision and context facilitated the demonstration and test bed approach to the ULL at Mueller, which may have foreclosed opportunities for more radical socio-technical innovations producing more just and sustainable outcomes.

3.1. Demonstrating the potential of urban smart grids

Providing vision and leadership is essential to governing in a democracy (Ezrahi, 1990). ULL test out competing visions of urban energy futures that align various actors around extensive reconfigurations of urban infrastructures (Bulkeley, Castan Broto and Maassen, 2013). These visions reflect broadly shared values and beliefs about technologies and their social impacts (Jasanoff, 2004; Jasanoff and Kim, 2009). ULL offer ways to produce, reinforce, and strengthen particular visions of technologically mediated urban futures (Bulkeley and Castán Broto, 2013b; Hodson and Marvin, 2009a; Reimer et al., 2012). At the same time, they provide opportunities to address global climate and energy concerns with particular "testable" or demonstrated solutions in localised places (Bulkeley and Castán Broto, 2012; Castán Broto and Bulkeley, 2013a). Alternative conceptions of urban experimentation offer different and often competing modes of knowledge production about urban sustainability that provide a different set of norms and rules by which communities can respond to climate change.

In the case of the Pecan Street smart grid demonstration project, it was cast both as a demonstration and a place of learning and experimentation. Given the context of the project – the Mueller development – the Environmental Defense Fund (2014) ardently promoted it:

> The Mueller neighborhood, the locus of Pecan Street, is a laboratory of ideas and technologies that will move the nation's $1.3 trillion electricity market toward a future in which energy is cheap, abundant and clean. If Pecan Street is successful, every neighborhood in America will look like it in 20 years.

Creating knowledge in ULL has an intense focus on learning and demonstration. Thus, the way people view the ULL and the knowledge they generate or demonstrate is integral to the activities conducted within. By focusing on energy monitoring and data collection, the Pecan Street database, analysis, and sharing wiki, Dataport, is a fundamental resource for energy research on smart grids. This database serves several purposes. First, it offers researchers data on energy usage where smart technologies have been implemented. This influences research findings and possibly, future policies regarding smart grid systems. Second, it offers companies opportunities to see how effective their technologies are, both in terms of energy efficiency or reliability, and in terms of customer acceptance. As one Pecan Street representative explained:

> We're always happy to exchange information with people. It really helps that we are a non-profit. I talk to cities; I talk to for-profit companies. We meet with them and they say, what have you learned, and I'll be happy to tell the for-profit company that is trying to build a product that this is what we've learned, this is what's failed, and this is what's succeeded. [. . .] My job is to make sure we can get as much data as possible to give to people so they can utilize it and learn from it.
>
> (Interview, October 2015)

ULL, thus, can be viewed as a "theatre of proof" (Simakova, 2010; Smith, 2009) for ways of configuring smart technologies in urban space to achieve sustainable, low-carbon outcomes. Stemming from work in STS around the role of demonstrations and public engagement with technology (Laurent, 2011; Marres, 2011; Marres and Lezaun, 2011; Rosental, 2014), especially in contemporary practices of the product "launch" in high-tech industries, this notion of the theatre of proof typically is framed by the situation where an organisation "offers a "novel" product to "the market" (Simakova, 2010, p. 549). ULL are not only places of knowledge production, but venues for linking technological artefacts and publics. In this sense, ULL serve as mediators between possible socio-technical futures and a wider public who might adopt the knowledge or technological systems emanating from the living lab.

3.2. ULL and test bed urbanism

Connecting local, place-based "experiments" to broader urban transitions has been the subject of much research on urban sustainability transitions. As protected niches, ULL might serve as mediators in urban energy transitions, but they are also geographical configurations that leverage local and regional assets to address more than local concerns and influence broader global audiences. The conception of test-bed urbanism (Halpern et al., 2013) helps to explain how urban spaces are configured for technological testing, as "platforms" for ICT and smart city technology development. The rise of so-called "platform capitalism" (Morozov, 2015) makes information infrastructure central to the provision of urban services, from transportation to energy, marked by the growth of the knowledge-based and sharing economies.

The platform metaphor is popular in the electricity industry. In Austin, Pecan Street's conception was formulated around the idea of an "energy Internet" – an open platform for testing a variety of smart grid technologies. The group developed their own open platform that fits with the broader movement towards creating an "information technology platform that makes possible a wide range of new products and services that provide customer value" where "mobile phone app stores and the Internet provide powerful examples of how a grid operator can earn more revenue and catalyse significant private sector opportunity by structuring its grid as a platform for a broad range of private sector activity" (Pecan Street, 2011, p. 7). The obvious allusions to the platform services such as Facebook, Amazon and Uber are discussed here as models for the electric grid.

This allusion to cell phone apps provides a vision of a radically decentralised electricity system where grid operators supply a platform and utilities and new energy companies offer energy services to customers in a highly competitive electric marketplace. This vision has captivated a whole array of private sector actors operating on speculative future scenarios to capture the growing smart grid and IoT markets. Cities operating in financially restricted positions with limited budgets and pressures to develop and meet a variety of public needs find this opportunity to attract capital enticing, thus creating entrepreneurial strategies to retain large smart grid and IoT companies.

In Austin, the Chamber of Commerce has a specific strategy to attract start-ups with potential to receive venture capital for growing their companies. Start-ups working on clean energy and power technology, creative and digital media technology or data management are able to add to the key industries in Austin, all of which relate to smart grid technology and development. The Austin Chamber of Commerce boasts the municipal utilities commitment to renewable power, the Pecan Street's research potential, and ERCOT (Electric Reliability Council of Texas)'s willingness to integrate clean energy companies into their electric grid as drawing points for energy companies. Supportive of these efforts is the University of Texas Clean Energy Incubator, the CleanTX cluster development organisation, and the already large clean tech industry located in Austin. But the city more largely is thought of as an experimental space for these companies, nurtured by the various resources the city offers. As one Chamber of Commerce representative explained:

> We were tasked with bringing the industry, bringing the clean industry. [. . .] What's the future of clean energy, clean tech? Its that efficiency piece, right. So, clean energy has grown into from renewables, natural gas, whatever into this "how can we do things better, cheaper, faster"? Pecan Street is a great example. [. . .] There are a lot of software engineers here, there are a lot of people who know how to analyze data. Austin is a good fit for those companies. This is a natural place for them to be. [Clean tech] is going towards devices that communicate to create efficiencies. Austin [has a] lot of software engineers, and there is an incredible quality of life. You've got the Pecan Street Project where companies can test their sensors.
>
> (Chamber of Commerce Representative, Interview,
> November 2015)

ULL seem to be far from neutral technological niches, but strategic resources for cities to attract capital. The generalised version of this approach to governance suggests that these spaces aren't merely "niches", but important parts of a strategy to transform the urban fabric into a platform for socio-technological experimentation. This implies that social practices and infrastructures are malleable rather than obdurate and structured. But this fails to account for social divisions and social structures that shape everyday practice. In Austin, for example having a large public–private redevelopment project provided the opportunity for the smart grid demonstration project to flourish in a community of so-called "early adopters": largely upper middle class residents that are motivated to save energy or participate in new technology testing.

> Like, these incubator companies will come up to us, I have this product, I need to get it field tested. We have three hundred volunteers in this neighborhood who will let us install it in their house. And, I would say, more than half of them want it in there, and the other half you usually have to convince a little bit with a financial incentive [. . .] and most of them will say yes, but more than half will jump at the fact to become a test-bed. So I would wholeheartedly agree with that synopsis of Mueller [as a living laboratory]. It's a really great test-bed of people that are, you know, early-adopters.
>
> (Pecan Street representative, Interview, October 2015)

This points towards an area for further research on governance experimentation: the role of citizens. While there are examples of grassroots approaches to urban transitions that promise greater democratic engagement (Blanchet, 2015; Seyfang and Haxeltine, 2012; Smith, Hargreaves, Hielscher, Martiskainen and Seyfang, 2016), this Austin ULL utilises a top-down approach where residents are encouraged to install technologies, receive incentives or benefits for doing so, and in return, participate in the research that monitors energy consumption and performance. This approach relies on "early-adopters" who are willing to participate in already designed programs, unlike a grassroots approach that focuses on collective visioning processes and collective ownership to govern energy system change. The focus of approaches taken in Pecan Street research, instead, is on individual energy consumption behaviours and technological efficiency, fitting with the existing regime of energy provision. Certainly, ULL are useful for understanding the technical limits of the smart grid; however, this approach contributes to the lock-in of particular pathways for smart grid development without broader consideration of the various concerns of citizens or with structural limitations to managing energy consumption and production.

In addition, governing by experiment in demonstrations and test-beds suggests that a particular governable subject already exists. However, in ULL and other urban experiments, urban citizens are often made to be the object of engagement

– the engaged customer, active participant, technology adopter. The literature on urban experiments has treated these urban socio-technical interventions as projects applied to existing urban landscapes and populations. However, as suggested with urban smart grid experiments, the governing of energy use is enacted through particular imagined subjects.

4. Making smart consumers and governing demand

ULL projects offer new opportunities for various actors to participate in urban energy and sustainability transitions. If ULL take on test-bed or demonstration approaches, what are the roles of participants and other actors? Do test-bed approaches suggest citizens are just consumers awaiting more sustainable technologies, or are they passive observers of demonstration projects, waiting to lend their approval for new technological solutions? As smart grid experiments proliferate, customer-utility relations are being reconfigured. The consumer, or household, is positioned as an engaged and active consumer, making decisions about energy consumption throughout the day. The smart grid enacts a set of relations and practices, both materially and discursively, wherein the conduct of end users is governed through "technologies of government" guided by particular political-economic rationalities. Smart grid experiments reconfigure familiar domains and categories – the household, the consumer – to govern how everyday practices are performed.

For example, smart grid demonstration projects have increasingly used the vocabulary of customer engagement and empowerment (Gangale, Mengolini and Onyeji, 2013). The customer moves beyond the role as a passive consumer and becomes an active participant in the electricity grid with new responsibilities, choices and opportunities (Naus, van Vliet and Hendriksen, 2015). Yet, as this discourse becomes pervasive, there is still little evidence that households are afforded autonomy or agency for engaging in energy transitions, with existing regimes playing dominant roles in shaping the implementation and standardisation of smart grid systems (Goulden, Bedwell, Rennick-Egglestone, Rodden and Spence, 2014).

The growth in attention to demand response, time-of-use pricing, and other "customer side" interventions have been celebrated by utilities and electricity providers as potential opportunities to shave or shift peak demand while increasing customer awareness and engagement. Yet, these practices rely on significant changes in energy consumption that have not been realised (Hargreaves Nye and Burgess, 2013). Underlying much of these programs is a conceptualisation of the end-user as a rational economic actor, or what Strengers (2013, p. 51) calls "resource man" – "a data-driven, information-hungry, technology-savvy home energy manager". These depicted end users are imagined as smart subjects, conscribed by social norms, expected to perform scripted uses for smart technologies with the encouragement to act rationally and responsibly.

4.1. Rationalities of experimentation

Governmentality helps us analyse how power operates beyond consensus or violence, linking technologies of the self (self-regulation) with technologies of domination (discipline), while also providing a linkage between the state apparatus and the constitution of the subject. Foucault posited that power was about "governing the forms of self-government, structuring and shaping the field of possible action of subjects" (Lemke, 2002, p. 50), or in other words, the "conduct of conduct".

Foucault also suggested that political rationality creates a discursive field wherein the exercise of power is made rational, "examining *how forms of rationality inscribe themselves in practices or systems of practices*, and what role they play within them, because it's true that "practices" don't exist without a certain regime of rationality" (Foucault, Burchell, Gordon and Miller, 1991, p. 79, italics added). Thus, if we take governmentality as a way to understand governing of urban energy systems, we must understand the way governing operates through the practices of those being governed (i.e. subjects), and the political rationalities informing these technologies of governance.

For example, Bulkeley, Powells and Bell (2016, p. 20) argue that governing energy use in the smart grid "works through the disposition of socio-material configurations through which conducts unfold and accompanying processes of normalising what constitute both acceptable and optimal forms of conduct", and that smart grids entail a specific governmental program that works through "recomposing the ways in which everyday practices are conducted". It is in this sense that the reworking of relations of consumption and production in smart grid experiments can be understood through the notion of governmentality.

Smart grid experiments, as a governmental program and specific locale (i.e. in a ULL), suggest a proper and optimal form of energy conduct for their participants. The idea of "self as enterprise" (McNay, 2009) is central to the implementation, political legitimacy and "success" of smart grid experiments. The promise of smart grids relies, in one part, on behavioural changes of users, expecting "smart users" to become active participants in the smart grid, performing their part as solar pioneers, eco-energy misers or flexible energy users adjusting consumption to the dynamics of a time-of-use rate structure. In this sense, smart grid experiments "success" presupposes (rational and individual) market actors who manage their everyday practices in a careful, calculative and reflexive way.

But as the experience in some of the households in Austin's smart grid experiment show, people do not necessarily act "rationally". A Pecan Street representative explained this point directly with a story of a multi-family tenant and research participant:

> We found crazy stuff occurring. You know, we showed up to one unit. And, the guy knocks on the door, our technician, and he's like "Oh I'm here to diagnose, there is a problem with our monitoring device," and the guy goes, "Oh yeah what's going on?" And he's like, "well it always shows

your oven's on", and the guy goes, "yeah, my oven is on," and our technician was like, "no no no no, we *show* that your oven is on" and the guy's like, "yeah yeah yeah, my oven's on." And it turns out this guy just like left his oven on all the time.

(Pecan Street Representative, Interview, October 2015)

Similarly, a representative from the Environmental Defense Fund explained that from his research in the smart grid experiment in Austin, and on energy more generally, people just don't think or care about energy enough to change their behaviour or their practices.

It's a very wonky subject, its not necessarily the most interesting conversation material for a lot of people so one barrier is just peoples interest levels. There is a statistic that is widely quoted that people think about their energy bills and electricity six minutes a year. For most people its not something that you choose to focus on. [. . .] Even if they don't think about it that much, they think about ways to save money, if something is a no-brainer, then you make that choice.

(EDF Representative, Interview, October 2015)

As energy researchers engage with smart grid users, they often seem to get dismayed by the irrationality of human energy decisions. As the quote above illustrates, the researchers involved with smart grid experiments understand that end users don't necessarily think about energy very often, but they still feel they can be persuaded economically. However, this logic is changing the nature of smart grid implementation. Lessons learned from early studies on energy efficiency impacts of smart metres and in-home displays suggest little evidence of sustained behaviour change (Hargreaves, Nye and Burgess, 2010; Hargreaves et al., 2013). Studies on voluntary demand response and time of use pricing have indicated that these options may work (Dyson, Borgeson, Tabone and Callaway, 2014; Muratori, Schuelke-Leech and Rizzoni, 2014), but the trend towards automating decision making to maximise energy and economic efficiency seem to be the dominant trend. As one EDF representative explained:

In terms of energy efficiency and smart grid and how they are related, its just sort of the next evolution. It's using machines and technology that doesn't have the human error element or the human interest level. You have these items programmed to be more efficient and at scale that will take a lot of the human element of being more efficient with energy out of the equation. [. . .] I mean, humans might want to act with the environment in mind, but they have their priorities and they have a lot of other things to do that day, and some things slip through the cracks, and if you want to be a good environmentalist but that's a low priority for you that can slip through the cracks and the technology can make it a lot easier.

(EDF Representative, Interview, October 2015)

While automation may provide energy and cost savings for end users, it also rationalises and normalises the deep integration of smart technologies in everyday life most probably without deliberation over end users concerns or values. Putting in place of a technological fix for the seeming inflexibility of energy demand and the irrationality of human behaviour frames these problems as purely technical ones. But, these problems are more than technical. Energy demand is structured by the rhythms and patterns of everyday life (Walker, 2014). Consumption is not for the sake of consumption, but rather for aiding in everyday practices shaped by social norms, habits, economic demands, and other conventions (Shove, Pantzar and Watson, 2012; Shove and Walker, 2014). The limitations of the techno-economic approach exemplified by the forms of experimentation discussed in this chapter explains, however, that the design and implementation of ULL – whether for smart grid experiments or other purposes – need to be questioned along axes of social and political concern. Who shapes the agenda and vision of a living laboratory, what are the planned outcomes and impacts, and who benefits? All of these questions require further consideration in the study of ULL.

5. Conclusion

In this chapter, I have highlighted how a specific ULL was constructed materially and discursively as a place of demonstration for public approval and a test bed for smart technologies. Although there are certainly positive benefits from smart grid implementation and demonstration, the role of citizens in determining or influencing the pathways to a smart energy future seem to be limited to very narrow realms of participation through consumption. I suggested that the approach of neoliberal governmentality helps explain how the actions of users of ULL are governed – proper forms of conduct that adhere to particular governmental rationalities described by techno-economic concerns. However, a contradictory trend towards automation and algorithmic, market-minded decision making – a technological fix – is progressing such that the "empowered" consumer no longer needs to act on information provision, they simply adopt technologies and enrol in programs, reinstituting a passive consumer role.

Therefore, while the current approaches to smart grid experiments rely on reshaping the "conduct of conduct", there is a trend towards automation and a techno-economic fix which vastly diminishes the potential for democratic engagement with planning or shaping energy systems. Just as smart city strategies that promise safer, healthier, more democrati, and sustainable cities, urban smart grid projects promise greater roles for consumers and place responsibility on citizens to facilitate change under a constrained environment and set of conditions. Yet, the agency of users is limited to consumption habits, preferences and technology adoption. As ULL for smart grid projects serve as demonstrations for particular pathways, there is risk that lock-in will occur without engagement of a broader public sphere (Verbong, Beemsterboer and Sengers, 2013). Experimentation must account for alternative visions to avoid furthering the notion that users and citizens are just barriers to smart grid implementation.

With the ever-greater entrenchment of smart technologies, greater amounts of data are also being collected and analysed. With the smart grid, this offers opportunities for the deepening of surveillance in everyday life (Klauser and Albrechtslund, 2014), while at the same time strengthening the opportunities for "corporate storytelling" to further normalise unequal social relations in the smart grid and the smart city by placing the private corporation – with profit motives – at the centre of the construction and implementation of smart urban technologies (Söderström, Paasche and Klauser, 2014). The governmental rationalities of these projects must also take into account these private interests and the prescribed roles for urban citizens – now as smart-users, Resource Men or neoliberal consumers.

By highlighting how particular governmental rationalities enable particular technologies to arise as solutions to urban problems, the problematics of "smart" user subjectivities in smart grid experiments, and the limitations of demonstration and test-bed approaches to ULL, this chapter has raised several critical issues for future scholarship on ULL. First, although ULL promise ways to test-out and experiment with solutions to climate change and urban sustainability, they are in large part shaped by existing socio-technical regimes with political economic interests and goals of developing technological fixes to urban problems. This may contribute to the creation of pockets of sustainable development in the city where only small portions of the population benefit. However, ULL together with bottom-up, citizen driven action can facilitate more radical and alternative changes, a point which should not be discounted.

Second, as demonstrations, ULL are significant opportunities to enrol public support for addressing key issues in transitions to urban sustainability or renewable energy. Yet, these approaches seem to have a limited conception of how users can interact with and participate in energy transitions. The Austin case study demonstrates that techno-economic approaches seek to regulate the conduct of individuals through economic incentives, but this limited approach fails when people act irrationally. Thus, a less democratic option of centralised automation and control is being pursued in smart grid implementation as a technological fix for the barriers of active human and user participation. These trends point towards the necessity of reinvigorating experimentation in ULL with a radically democratic agenda. We should take seriously the role that seemingly one-off experiments have for possible co-production of more sustainable and just urban futures.

References

Alahakoon, D. and Yu, X. (2013). 'Advanced analytics for harnessing the power of smart meter big data'. *2013 IEEE International Workshop on Inteligent Energy Systems (IWIES)*, 40–45. Retrieved from https://doi.org/10.1109/IWIES.2013.6698559

Allen, B. L. (2011). 'Laboratorization and the "Green" rebuilding of New Orleans's lower ninth ward'. In C. Johnson (Ed.), *The neoliberal deluge: Hurricane Katrina, late capitalism, and the remaking of New Orleans*. Minneapolis, MN: University of Minnesota Press.

Barry, A. (2006). 'Technological zones'. *European Journal of Social Theory*, 9(2), 239–253. Retrieved from https://doi.org/10.1177/1368431006063343

Bijker, W. E., Hughes, T. P. and Pinch, T. J. (1987). *The social construction of technological systems: New directions in the sociology and history of technology.* Cambridge, MA: MIT Press.

Blanchet, T. (2015). 'Struggle over energy transition in Berlin: How do grassroots initiatives affect local energy policy-making?' *Energy Policy, 78,* 246–254. Retrieved from https://doi.org/10.1016/j.enpol.2014.11.001

Bulkeley, H. and Castán Broto, V. (2012). 'Urban experiments and climate change: Securing zero carbon development in Bangalore'. *Contemporary Social Science, May 2014,* 1–22. Retrieved from https://doi.org/10.1080/21582041.2012.692483

Bulkeley, H. and Castán Broto, V. (2013a). 'A survey of urban climate change experiments in 100 cities'. *23,* 92–102. Retrieved from https://doi.org/10.1016/j.gloenvcha.2012.07.005

Bulkeley, H. and Castán Broto, V. (2013b). 'Government by experiment? Global cities and the governing of climate change'. *Transactions of the Institute of British Geographers, 38*(3), 361–375. Retrieved from https://doi.org/10.1111/j.1475-5661.2012.00535.x

Bulkeley, H., Castan Broto, V. and Maassen, A. (2013). 'Low-carbon transitions and the reconfiguration of Urban Infrastructure'. *Urban Studies, 51*(7), 1471–1486. Retrieved from https://doi.org/10.1177/0042098013500089

Bulkeley, H., Powells, G. and Bell, S. (2016). 'Smart grids and the constitution of solar electricity conduct'. *Environment and Planning A, 48*(1), 7–23. Retrieved from https://doi.org/10.1177/0308518X15596748

Carvallo, A. and Cooper, J. (2015). *The Advanced Smart Grid: Edge Power Driving Sustainability* (2nd ed.). Boston, MA: Artech House.

Castán Broto, V. and Bulkeley, H. (2013a). 'A survey of urban climate change experiments in 100 cities'. *Global Environmental Change: Human and Policy Dimensions, 23*(1), 92–102. Retrieved from https://doi.org/10.1016/j.gloenvcha.2012.07.005

Castán Broto, V. and Bulkeley, H. (2013b). 'Maintaining climate change experiments: Urban political ecology and the everyday reconfiguration of urban infrastructure'. *International Journal of Urban and Regional Research, 37*(6), 1934–1948. Retrieved from https://doi.org/10.1111/1468-2427.12050

Clark-Madison, M. (2002, January 25). *Who Will Rule Mueller?* Retrieved from www.austinchronicle.com/news/2002-01-25/84459/

Davidson, K. and Gleeson, B. (2014). 'The sustainability of an entrepreneurial city?' *International Planning Studies, April,* 1–19. Retrieved from https://doi.org/10.1080/13563475.2014.880334

Dyson, M. E. H., Borgeson, S. D., Tabone, M. D. and Callaway, D. S. (2014). 'Using smart meter data to estimate demand response potential, with application to solar energy integration'. *Energy Policy, 73,* 607–619. Retrieved from https://doi.org/10.1016/j.enpol.2014.05.053

Evans, J. and Karvonen, A. (2013). ' "Give me a laboratory and I will lower your carbon footprint!" Urban laboratories and the governance of low-carbon futures'. *International Journal of Urban and Regional Research, 38,* 413–430. Retrieved from https://doi.org/10.1111/1468-2427.12077

Evans, J. and König, A. (2013). 'Introduction: Experimenting for sustainable development? Living laboratories, social learning and the role of the university'. In A. König (Ed.), *Regenerative sustainable development of universities and cities* (pp. 1–24). Cheltenham, UK, and Northampton, MA: Edward Elgar Publishing.

Ezrahi, Y. (1990). *The descent of Icarus: Science and the transformation of contemporary democracy.* Cambridge, MA: Harvard University Press.

Foucault, M., Burchell, G., Gordon, C. and Miller, P. (1991). *The Foucault effect: Studies in governmentality.* Chicago, IL: University of Chicago Press.

Frangoul, A. (2015). Pecan Street, Inc. | Is this the world's most innovative neighborhood? Retrieved from www.pecanstreet.org/2015/03/is-this-the-worlds-most-innovative-neighborhood/

Gangale, F., Mengolini, A. and Onyeji, I. (2013). 'Consumer engagement: An insight from smart grid projects in Europe'. *Energy Policy, 60*, 621–628. Retrieved from https://doi.org/10.1016/j.enpol.2013.05.031

Geels, F. W. (2007). 'Transformations of large technical systems: A multilevel analysis of the Dutch highway system (1950–2000)'. *Science, Technology and Human Values, 32*(2), 123–149. Retrieved from https://doi.org/10.1177/0162243906293883

Geels, F. W. and Kemp, R. (2007). 'Dynamics in socio-technical systems: Typology of change processes and contrasting case studies'. *Technology in Society, 29*(4), 441–455. Retrieved from https://doi.org/10.1016/j.techsoc.2007.08.009

Gieryn, T. F. (2006). 'City as truth-spot: Laboratories and field-sites in urban studies'. *Social Studies of Science, 36*(1), 5–38. Retrieved from https://doi.org/10.1177/0306312705054526

Gopakumar, G. (2014). 'Experiments and counter-experiments in the urban laboratory of water-supply partnerships in India'. *International Journal of Urban and Regional Research, 38*(2), 393–412. Retrieved from https://doi.org/10.1111/1468-2427.12076

Goulden, M., Bedwell, B., Rennick-Egglestone, S., Rodden, T. and Spence, A. (2014). 'Smart grids, smart users? The role of the user in demand side management'. *Energy Research and Social Science, 2*, 21–29. Retrieved from https://doi.org/10.1016/j.erss.2014.04.008

Graham, S. and Marvin, S. (2001). *Splintering urbanism: networked infrastructures, technological mobilities and the urban condition.* London and New York: Routledge.

Hall, T. and Hubbard, P. (1996). 'The entrepreneurial city: new urban politics, new urban geographies?' *Progress in Human Geography, 20*(2), 153–174. Retrieved from https://doi.org/10.1177/030913259602000201

Halpern, O., LeCavalier, J., Calvillo, N. and Pietsch, W. (2013). 'Test-bed urbanism'. *Public Culture, 25*(270), 272–306. Retrieved from https://doi.org/10.1215/08992363-2020602

Hargreaves, T., Nye, M. and Burgess, J. (2010). 'Making energy visible: A qualitative field study of how householders interact with feedback from smart energy monitors'. *Energy Policy, 38*(10), 6111–6119. Retrieved from https://doi.org/10.1016/j.enpol.2010.05.068

Hargreaves, T., Nye, M. and Burgess, J. (2013). 'Keeping energy visible? Exploring how householders interact with feedback from smart energy monitors in the longer term'. *Energy Policy, 52*, 126–134. Retrieved from https://doi.org/10.1016/j.enpol.2012.03.027

Harvey, D. (1989). 'From managerialism to entrepreneurialism: The transformation in urban governance in late capitalism'. *Geografiska Annaler, 71*(1), 3–17.

Heathcott, J. (2005). '"The whole city is our laboratory": Harland Bartholomew and the production of urban knowledge'. *Journal of Planning History, 4*(4), 322–355. Retrieved from https://doi.org/10.1177/1538513205282131

Hodson, M. and Marvin, S. (2009a). 'Cities mediating technological transitions: Understanding visions, intermediation and consequences'. *Technology Analysis and Strategic Management, 21*(4), 515–534. Retrieved from https://doi.org/10.1080/09537320902819213

Hodson, M. and Marvin, S. (2009b). '"Urban ecological security": A New Urban Paradigm?' *International Journal of Urban and Regional Research, 33*(1), 193–215. Retrieved from https://doi.org/10.1111/j.1468-2427.2009.00832.x

Hodson, M. and Marvin, S. (2010). 'Urbanism in the anthropocene: Ecological urbanism or premium ecological enclaves?' *City, 14*(3), 298–313. Retrieved from https://doi.org/10.1080/13604813.2010.482277

Jasanoff, S. (2004). *States of knowledge* (S. Jasanoff, Ed.). London: Routledge. Retrieved from https://doi.org/10.4324/9780203413845

Jasanoff, S. and Kim, S.-H. (2009). 'Containing the atom: Sociotechnical imaginaries and nuclear power in the United States and South Korea'. *Minerva, 47*(2), 119–146. Retrieved from https://doi.org/10.1007/s11024-009-9124-4

Karvonen, A. and van Heur, B. (2014). 'Urban laboratories: Experiments in reworking cities'. *International Journal of Urban and Regional Research, 38*(2), 379–392. Retrieved from https://doi.org/10.1111/1468-2427.12075

King, M. (2015, November 6). Mueller vs. TCAD: No room for affordability? Retrieved from www.austinchronicle.com/news/2015-11-06/mueller-vs-tcad-no-room-for-affordability/

Klauser, F. R. and Albrechtslund, A. (2014). 'From self-tracking to smart urban infra-structures: Towards an interdisciplinary research agenda on Big Data'. *Surveillance and Society, 12*(2), 273–286.

Knorr-Cetina, K. (1995). 'Laboratory studies: The cultural approach to the study of science'. In S. Jasanoff, G. E. Markle, J. C. Peterson and T. Pinch (Eds.), *Handbook of science and technology studies* (pp. 141–166). Thousand Oaks, CA: Sage Publications, Inc.

Laurent, B. (2011). 'Technologies of democracy: Experiments and demonstrations'. *Science and Engineering Ethics, 17*(4), 649–66. Retrieved from https://doi.org/10.1007/s11948-011-9303-1

Lemke, T. (2002). 'Foucault, governmentality, and critique'. *Rethinking Marxism, 14*(3), 49–64. Retrieved from https://doi.org/10.1080/089356902101242288

Livingstone, D. N. (2010). *Putting science in its place: Geographies of scientific knowledge.* Chicago, IL: University of Chicago Press.

Long, J. (2016). 'Constructing the narrative of the sustainability fix: Sustainability, social justice and representation in Austin, TX'. *Urban Studies, 53*(1), 149–172. Retrieved from https://doi.org/10.1177/0042098014560501

Luque, A. (2014). 'The smart grid and the interface between energy, ICT, and the city'. In T. Dixon, M. Eames, M. Hunt and S. Lannon (Eds.), *Urban retrofitting for sustainability: Mapping the transition to 2050.* New York: Routledge.

MacLeod, G. (2002). 'From urban entrepreneurialism to a 'Revanchist City?' On the spatial injustices of Glasgow's renaissance'. *Antipode, 34*(3), 602–624. Retrieved from https://doi.org/10.1111/1467-8330.00256

Marres, N. (2011). 'The costs of public involvement: Everyday devices of carbon accounting and the materialization of participation'. *Economy and Society, 40*(4), 510–533. Retrieved from https://doi.org/10.1080/03085147.2011.602294

Marres, N. and Lezaun, J. (2011). 'Materials and devices of the public: an introduction'. *Economy and Society, 40*(4), 489–509. Retrieved from https://doi.org/10.1080/03085147.2011.602293

Marvin, S., Luque-Ayala, A. and McFarlane, C. (2015). *Smart urbanism: Utopian vision or false dawn?* London and New York: Routledge.

McLean, A., Bulkeley, H. and Crang, M. (2015). 'Negotiating the urban smart grid: Socio-technical experimentation in the city of Austin'. *Urban Studies.* Retrieved from https://doi.org/10.1177/0042098015612984

McNay, L. (2009). 'Self as enterprise dilemmas of control and resistance in Foucault's the birth of biopolitics'. *Theory, Culture and Society, 26*(6), 55–77. Retrieved from https://doi.org/10.1177/0263276409347697

Morozov, E. (2015, June 6). 'Where Uber and Amazon rule: Welcome to the world of the platform'. *The Guardian.* Retrieved from www.theguardian.com/technology/2015/jun/07/facebook-uber-amazon-platform-economy

Muratori, M., Schuelke-Leech, B.-A. and Rizzoni, G. (2014). 'Role of residential demand response in modern electricity markets'. *Renewable and Sustainable Energy Reviews, 33,* 546–553. Retrieved from https://doi.org/10.1016/j.rser.2014.02.027

Naus, J., van Vliet, B. J. M. and Hendriksen, A. (2015). 'Households as change agents in a Dutch smart energy transition: On power, privacy and participation'. *Energy Research and Social Science, 9,* 125–136. Retrieved from https://doi.org/10.1016/j.erss.2015.08.025

Oudshoorn, N. and Pinch, T. J. (2005). *How users matter: The co-construction of users and technology.* Cambridge, MA: MIT Press.

Pecan Street. (2011). *Pecan Street Project Energy Internet Demonstration Request for Information.* Retrieved from www.pecanstreet.org/wordpress/wp-content/uploads/2011/02/RFI-Pecan_St_Project1.pdf

Ranganathan, P. and Nygard, K. (2011). 'Smart grid data analytics for decision support'. *2011 IEEE Electrical Power and Energy Conference,* 315–321. Retrieved from https://doi.org/10.1109/EPEC.2011.6070218

Reeves, K. (2008, October 3). 'Mueller revisits affordability and density'. Retrieved from www.austinchronicle.com/news/2008-10-03/681440/

Reimer, M. H., Mccormick, K., Nilsson, E. and Arsenault, N. (2012, August). 'Advancing sustainable urban transformation through living labs: Looking to the Öresund region'. Paper presented at the international conference on sustainability transitions.

Rhodes, J. D., Upshaw, C. R., Harris, C. B., Meehan, C. M., Walling, D. A., Navrátil, P. A., . . . Webber, M. E. (2014). 'Experimental and data collection methods for a large-scale smart grid deployment: Methods and first results'. *Energy, 65,* 462–471. Retrieved from https://doi.org/10.1016/j.energy.2013.11.004

Rosental, C. (2014). 'Toward a sociology of public demonstrations'. *Sociological Theory, 31*(4), 343–365. Retrieved from https://doi.org/10.1177/0735275113513454

Seyfang, G. and Haxeltine, A. (2012). 'Growing grassroots innovations: Exploring the role of community-based initiatives in governing sustainable energy transitions'. *Environment and Planning C-Government and Policy, 30*(3), 381–400. Retrieved from https://doi.org/10.1068/c10222

Shove, E., Pantzar, M. and Watson, M. (2012). *The dynamics of social practice: Everyday Life and how it Changes.* London: SAGE Publications.

Shove, E. and Walker, G. (2014). 'What is energy for? Social practice and energy demand'. *Theory, Culture and Society, 31*(5), 41–58.

Simakova, E. (2010). 'RFID "Theatre of the proof": Product launch and technology demonstration as corporate practices'. *Social Studies of Science, 40*(4), 549–576. Retrieved from https://doi.org/10.1177/0306312710365587

Smith, A., Hargreaves, T., Hielscher, S., Martiskainen, M. and Seyfang, G. (2016). 'Making the most of community energies: Three perspectives on grassroots innovation'. *Environment and Planning A, 48*(2), 407–432. Retrieved from https://doi.org/10.1177/0308518X15597908

Smith, W. (2009). 'Theatre of use: A frame analysis of information technology demonstrations'. *Social Studies of Science, 39*(3), 449–480. Retrieved from https://doi.org/10.1177/0306312708101978

Söderström, O., Paasche, T. and Klauser, F. (2014). 'Smart cities as corporate storytelling'. *City, 18*(3), 307–320. Retrieved from https://doi.org/10.1080/13604813.2014.906716

Strebel, I. and Jacobs, J. M. (2014). 'Houses of experiment: Modern housing and the will to laboratorization'. *International Journal of Urban and Regional Research.* Retrieved from https://doi.org/10.1111/1468-2427.12079

Strengers, Y. (2013). *Smart energy technologies in everyday life: Smart Utopia?* Basingstoke, UK: Palgrave Macmillan.

Swyngedouw, E. (2011). 'The non-political politics of climate change'. *ACME: An International Journal for Critical Geographies*, 12(1), 1–8.

Verbong, G. P. J., Beemsterboer, S. and Sengers, F. (2013). 'Smart grids or smart users? Involving users in developing a low carbon electricity economy'. *Energy Policy, 52*, 117–125. Retrieved from https://doi.org/10.1016/j.enpol.2012.05.003

Voytenko, Y., McCormick, K., Evans, J. and Schliwa, G. (2016). 'Urban living labs for sustainability and low carbon cities in Europe: Towards a research agenda'. *Journal of Cleaner Production, 123*, 45–54. Retrieved from https://doi.org/10.1016/j.jclepro.2015.08.053

Walker, G. (2014). 'The dynamics of energy demand: Change, rhythm and synchronicity'. *Energy Research and Social Science, 1*, 49–55. Retrieved from https://doi.org/10.1016/j.erss.2014.03.012

While, A., Jonas, A. E. G. and Gibbs, D. (2004). 'The environment and the entrepreneurial city: Searching for the urban sustainability fix in Manchester and Leeds'. *International Journal of Urban and Regional Research*, 28(3), 549–569.

While, A., Jonas, A. E. G. and Gibbs, D. (2010). 'From sustainable development to carbon control: Eco-state restructuring and the politics of urban and regional development'. *Transactions of the Institute of British Geographers, 35*(1), 76–93. Retrieved from https://doi.org/10.1111/j.1475-5661.2009.00362.x

5

SMART CITY CONSTRUCTION

Towards an analytical framework for smart urban living labs

Frans Sengers, Philipp Späth and Rob Raven

1. Introduction

Cities are back on the international agenda as key-sites for negotiating and shaping sustainable development, economic growth, technological innovation, social cohesion and the like. As urban actors are increasingly confident about their potential and roles in transforming cities, a new wave of "experimental governance" is emerging, which gives centre stage to an actionable form of governance in "urban living labs" or "urban experiments" (Bulkeley and Castán Broto, 2013; Evans, Karvonen and Raven, 2016; Sengers et al., 2016; Voytenko et al., 2015; Wolfram and Frantzeskaki, 2016).

In our view, an analytical distinction should be made between the notions of experimentation, experiments and (urban) living labs. Experiments are concrete "hands-on" individual initiatives (see Sengers et al., 2016 for a definition) while an (urban) living lab can be conceptualised as a bounded site where multiple experiments take place or otherwise as the institutional aggregation of multiple experiments (see Voytenko et al., 2016 for a recent overview). Experimentation is the overarching term referring to the act of conducting individual experiments and setting up and running urban living labs as well as providing broader support to such ventures. We adhere to this strict analytical distinction even though most academics and practitioners involved in experimentation often use terms interchangeably.

Academic scholarship in urban studies and transition studies has started to explore the ways in which experimentation may shape wider processes of urban transformation, but little explicit attention has been paid to "smart" urban living labs and the set of experiments conducted in these lab sites. This is notable, because in recent years, the *Smart City* has emerged as a highly popular term among engineers, policy makers, architects, scholars and others interested in how the growth of connected information and communication technologies (ICTs) might reshape the social and material fabric of cities, including their sustainability (Hajer and Dassen, 2014; Hodson and Marvin, 2014). Scholars continue to struggle how to make sense of what a smart city is and what kind of political, social and material implications smart cities may generate (Luque-Ayala and Marvin, 2015; Marvin et al., 2016). Notably, the emerging literature on urban experimentation has not yet engaged explicitly with this relatively young term in urban development, despite that it has quickly approached and even surpassed popularity of previous terms, such as eco-cities or sustainable cities, for articulating urban futures (De Jong et al., 2015; Evans, Karvonen and Raven, 2016).

The aim of this chapter is to lay the grounds for a fresh and multi-faceted perspective on how the smart city of the future is being *constructed* today. We use the term "construction" both in the literal and the figurative sense. On the one hand, many of the actors involved are eager to start with the physical act of construction, erecting new ICT-integrated buildings, enmeshing steel and concrete with sensors and fibre optics. On the other hand, we realise that new technologies and urban utopias are also "socially constructed", highlighting that technological development is neither autonomous nor deterministic, but a quirky and contingent process deeply shaped by human action and by the sometimes-surprising outcomes of place-specific activities in urban living labs (Bijker et al., 1987; Latour, 1996). Although the focus of this chapter is thus on smart city experimentation, we believe that the framework that we develop has broader potential to the study of urban experiments and urban living labs. In this way, we are responding to recent calls for further analysis into the ways in which urban living labs may become embedded into (or otherwise interact with) incumbent modes of governance and institutions (Voytenko et al., 2016). As such, this chapter speaks to the first set of questions and addressed in this edited volume, in particular to the second question on the variety of principles and modes of governance that are being developed through experimentation.

In this chapter, we develop a *socio-technical perspective* that seeks to understand the smart city not as a fixed narrative or static artefact that simply diffuses throughout global cities, but as an emerging practice in which a range of public and private actors are trying to "actualise" smart urbanism. Such a perspective needs to take into account the intertwined and co-evolutionary relationship between technological development fuelled by innovative material artefacts on the one hand, and social processes such as new forms of governance, institutions and discourses on the other. While some recent contributions point to certain relevant elements (e.g. Carvalho, 2014; Luque-Ayala and Marvin 2015; Marvin et al., 2016),[1] a comprehensive socio-technical perspective on smart city experimentation is lacking as of yet.

To sketch out the contours of such socio-technical perspective, we mobilise a strand of academic literature, which can be seen as an offshoot of the earlier work on social construction of technology and evolutionary perspectives on innovation. The field of *sustainability transitions* has developed a substantial conceptual vocabulary to analyse major shifts in socio-technical systems tied up with the trajectories of urban development (Bulkeley et al., 2011; Markard et al., 2012; Nevens et al., 2013; Rutherford and Coutard, 2014). Cities, too, can be conceptualised as socio-technical systems or, more precisely, as localised patchworks of a range of socio-technical systems around the provision and use of mobility, energy, housing, recreation and the like, aligned through regimes of urban planning and governance. Smart city experiments and living labs can be considered as key arenas where transitions towards smart urbanism are being shaped and/or contested. The field of sustainability transitions has developed a substantial set of concepts and ideas on the role of experimentation in shaping systemic changes and, increasingly, urban transformations (Sengers et al., 2016). What, this paper asks, can this field offer to understand and make sense of smart city experimentation and its potential political, social and material implications? Addressing this question will also provide a fresh perspective on the analysis of urban living labs more generally.

The remainder of this chapter is structured as follows. Section 2 introduces the field of sustainability transitions and provides the main thrust of our argument. We argue that smart city experiments should be interpreted as simultaneous effort of discursive, institutional and material construction, and we show that cities are the sites in which actors experiment in various ways with each of these "constructions-in-action". In order to highlight current patterns and capture the diversity of on-going smart city developments, Section 3 (materiality), Section 4 (discourses) and Section 5 (institutions) elaborate our conceptual argument and provide empirical illustration by using examples from Amsterdam, the Netherlands and Hamburg, Germany. Section 6 summarises our main argument, reflects on the implications and discusses how we can take this forward in terms of further conceptualisation and promising avenues for future research.

2. Insights from sustainability transitions: Discourses, institutions and materiality to investigate socio-technical experimentation

Scholars in the field of *sustainability transitions* investigate societal transformations toward sustainable socio-technical systems of production and consumption (Grin et al., 2010; Markard et al., 2012). The study of socio-technical transitions to sustainability draws on a wide range of literatures and lines of thought (such as neo-institutional theory, evolutionary economics and the social constructivist tradition within science and technology studies) and a variety of frameworks and approaches have been developed to express how sustainable alternatives can be empowered in the face of unsustainable socio-technical systems. We argue that two key elements stand out as the core of a transitions perspective.

One defining element that sets transitions thinking apart from the wider literature of social change and policy theory is its engagement with the process of *socio-technical experimentation* (Meadowcroft 2011; Van den Bergh, 2012). This is also relevant to more recent scholarly work on urban living labs (Bulkeley et al., 2015). Inspired by earlier STS work on how ideas and practices of natural science laboratories spill over to the wider world, the introduction of alternative technologies and practices can be seen as a messy experimental process that co-evolves with social and material realities. As such, the notion of experimentation in transition studies has a normative orientation: experiments are seen as important seeds of change that may eventually lead to a desirable and profound shift in the way a societal function – such as the provision of energy, water or mobility, or the urban experience for that matter – is being met. As precious yet-to-germinate microcosms of sustainable systems and practices, the alternative socio-technical configurations embodied in experiments offers a range of actors including users, citizens, policy makers and entrepreneurs to apply and test in real-life contexts a novel socio-technical configuration, with the aim of technological, social and institutional learning. The promise is that learning and demonstration effects of experiments add to the momentum of emerging sustainable configurations, which are geared to transform unsustainable socio-technical systems (Sengers et al., 2016).[2]

A second element that goes to the core of what defines a transition perspective is an engagement with the power struggle between *the forces of change versus the forces of stability*. While the earlier transitions work foregrounded the process of experimentation and the build-up of alterative systems, more recent work also stressed the process of destabilisation and the breakdown or reconfiguration of incumbent systems (Smith and Raven, 2012; Smink et al., 2015; Stegmaier et al., 2014; Turnheim and Geels, 2013). In order to get at the politics at play, a few recent contributions are relevant in distinguishing important conceptual categories that characterise this struggle.

Florian Kern has argued that explanations of transformative change are often based on one or more of the following three conceptual dimensions: interests, ideas and institutions (Kern, 2011).[3] Looking through the first lens of "interests", actors are seen as rational agents that exercise power over other agents to maximise their own utility. The notion of utility is expressed here in terms of "material wealth", so we could say that it qualifies as a perspective that incorporates material elements or notions of materiality. Looking through the second lens of "ideas", politics can be understood as a struggle for power played out through arguments about the "best story". Discourse analysis is proposed as a productive way to get at these storylines and how they frame problems and trigger change (Hajer, 2006). Because not the ideas as such but the way these ideas are articulated as storylines is emphasised, this qualifies as a discursive perspective. Looking through the third lens of "institutions", formal and informal rule sets embedded in organisational practice take centre stage. These vested regulative, normative and cognitive structures enable and constrain the political action of governments and other actors according to this institutional perspective (Scott, 1995).

A similar set of dimensions can be found in Frank Geels" recent work on how incumbents use power to resist transformative change (Geels, 2014). This includes material and discursive strategies as well as the notion of broader institutional power.[4] His first element here is "material strategy", which points to the need of actors to mobilise technical capabilities and financial resources to improve or optimise the performance of incumbent technologies. Incumbents are often well endowed in this respect as compared to newcomer firms or civil society actors. His second element is "discursive strategy", which points to the articulation of narratives that become dominant in setting the agenda by determining which issues are discussed how these are discussed. It is argued that the "discourse coalition" of policy makers and incumbent firms is often particularly powerful in this respect, especially when compared to the limited power alternative storylines articulated by, for instance, citizens groups or radical innovators (Hajer, 1995; Lindblom, 2001). Geels' third element is "broader institutional power", which points to power as embedded in political cultures, ideology and governance structures. It alludes to a post-political technocratic style of governance that facilitates the incremental strategies of incumbents and makes it difficult for newcomer firms or civil society actors to empower radically alternative transition pathways.

The point here is that categorisations of Geels and Kern both express a clear distinction between (1) *material*, (2) *discursive* and (3) *institutional* elements in characterising the struggle between stability and change of socio-technical systems. On the one hand, we understand these to form structural (enabling and constraining) contexts through which actors have to navigate when experimenting with new socio-technical configurations. On the other hand, material, discursive and institutional elements can form the actual object of experimentation, for example when actors trial out new technologies or new forms of collaborations between actors. As such, a way to study how this struggle unfolds in practice – albeit in a smaller scale – is to investigate what happens around today's real-life experiments and urban living labs with alternative technologies and practices geared to bring into being tomorrow's smart city. In the next sections, we elaborate on this threefold distinction described above as well as the idea of socio-technical experimentation.

3. Smart city experimentation as material site

The term smart city generally refers to the deep integration of ICTs in the urban fabric as a way to stimulate economic development (rendering a city competitive by boosting technological innovation and entrepreneurship) and to augment urban management (regulating a city in real-time by using big data, ubiquitous monitoring and predictive algorithms – see Kitchin, 2014). From a material perspective, the smart city can be conceptualised as an actual city that looks and feels different from the cities experienced by people today; a place where sensors, wires, plastic, glass, steel and concrete are put to task of measuring, regulating and managing the urban environment, and where flows of data channel through ICT infrastructures to the screens of analysts and decision makers in municipalities and businesses.

While the smart city is sometimes portrayed as an abstract utopia or dystopia to be realised in a faraway future, it already finds expression in the urban infrastructure and the buildings constructed today. The most famous smart city projects that pop to mind here are exceptional places such as Songdo, PlanIT

Valley and Masdar. These "greenfield sites" are presented as clean slates that seem to have no history, only a single predetermined future.[5] But the emerging material contours of the smart city can also be found in many urban areas throughout the world, thriving in vibrant inner-city cores as well as more peripheral and mundane places. These existing cities and "brownfield sites" are places with a messy history and multiple potential futures.

The first category of famous canonical experiences has been labelled "the smart city from scratch" (Carvalho, 2014) and the latter category of more mundane experiences of introducing of ICTs to reshape the material fabric of the urban has been labelled "the actually existing smart city" (Shelton et al., 2015). The material construction of the canonical smart city from scratch is sometimes envisioned by planners and engineers as simply the top-down "rollout" of next-generation high-tech infrastructure concentrated in one previously empty place. But for the material construction of the mundane actually-existing-smart-city and the constraints and opportunities imposed by history and the obdurate legacy of the built environment come into play (Hommels, 2005). Actors are forced to deal with a multitude of voices to test various technological elements of what a smart city might become and negotiate the piecemeal retrofit and installation of new technologies in the existing building stock located on a patchwork of disconnected experimental sites. This provides a different challenge altogether, but one that is more reminiscent of the situation in most parts of the world, including in Amsterdam and Hamburg, to which we will now turn.

Although most smart city ambitions of Amsterdam and Hamburg have not yet materialised, in some places these ambitions have become a material reality by being translated into a number of living labs populated by clusters of experiments. Such experiments change the physical constitution of the urban environment and are geared to re-shape the cityscape, building stock, material infrastructure and the spatial layout. Most of these living labs are located in different strategically chosen places throughout town. In Hamburg, where some resistance against the smartening of transportation infrastructures can be expected due to privacy concerns of citizens, experimentation has been started in the privately owned harbour area, and is planned to be expanded from there. Also the full rollout of smart metering to all customers – irrespective of their annual electricity consumption – is tested in Hamburg's new-built district HafenCity. In Amsterdam, a football stadium and surrounding business park as well as an old industrial area and a former navy yard were chosen as living labs for new configurations. Around 100 experimental initiatives are conducted in these three living lab areas or in other strategic locations throughout Amsterdam (ASC, 2016). In one of these designated living lab areas, the Buiksloterham neighborhood, a local architect explained that this is

> an experimental area . . . we're just doing it, that's the main starting point. We don't get bogged down in thinking about the organization and getting all permissions . . . that gives so much energy and very much spin-off with other parties to really make a difference in this place.

This illustrates the power of this kind of "hands-on" approach of conducting material experiments as transforming a part of the city physically, but also as an agent of institutional change and as mobilising force to rally an even stronger coalition of actors to enable even more material changes.

4. Smart city experimentation as discursive arena

The smart city is not only an actual material entity, but also one of the many contending buzzwords that reflect ideas on future urban life. Many such terms have entered policy and academic discourse in recent years: eco-cities, resilient cities, circular cities, low-carbon cities, creative cities. The smart city is the heir to terms such as the information city, the digital city, the intelligent city, the city of bits and smart urban management (Hollands, 2008; Söderström et al., 2014; Vanolo, 2014). In recent years, the term was popularised and heavily promoted by hi-tech companies like IBM and Cisco. It has completely eclipsed its associates and predecessors and – more strikingly – its use is growing much faster than other green and sustainable city terms, making it a particularly booming buzzword (De Jong et al., 2015).

As in any debate, various actors are projecting their interpretation onto this potentially powerful notion, vying over its meanings and implications. Where some see brink of a technological utopia (efficient resource management and new opportunities for democratic participation), others see doom and gloom (a new corporate agenda that leads to exacerbated forms of social exclusion and inequality) and they contest ways in which the impending thrust for smartness will reshape humanity's life cities – for better or for worse (Campbell, 2012; Deakin and Al Waer, 2012; Greenfield, 2013; Hollands, 2008; Marvin et al., 2016). From a discursive perspective, the smart city is a feat of "corporate storytelling" about an ICT-efied urban future; a narrative of a city that is seamless, connected, safe and manageable flanked by a counter narrative of a city that is inhuman, splintered, intrusive and uncontrollable. Smart urbanism hence is a discursive shift that can be traced in debates about urban futures and infrastructure development, with new subjects being created and contested, and a struggle over discursive hegemony (Hajer, 2006; Hajer and Dassen, 2014; Söderström et al., 2014; Vanolo, 2014).

One striking element in the rhetoric that perpetuates the smart city label and enables certain practices in the allegedly unstoppable rise of modernity is the explicit call by practitioners and policy makers to conduct experiments and to set up living labs. In Hamburg, for instance, the lord mayor wants his city to develop "into a laboratory of digital modernity" (Stadt Hamburg, 2016). In 2014, the administration signed a memorandum of understanding with the infrastructure provider CISCO to jointly develop a number of pilot studies of smart infrastructure. On the press conference, the lord mayor portrayed the trend towards smart city management as something unstoppable. The municipality should therefore get involved with this trend already in the experimentation phase, in order to co-shape it, representing the public interest (Meinecke, 2016). This framing, however, is bluntly dominated by the perspective of the administration and one particular company. In contrast with the emphasis that (urban) living lab literatures put on the role of citizen involvement, comparatively little effort has been made to rhetorically open up a debate with other stakeholders or citizens about what kind of smart infrastructure would be required to achieve the agreed upon objectives of urban development in Hamburg.

On the other hand, in Amsterdam ample lip service is provided to participation and the smart city "from the bottom up". When speaking at the annual smart city fair last year, the mayor of Amsterdam gave a speech that positioned Amsterdam as a "front runner smart city" with seamless links to the past:

> Amsterdam has a centuries old tradition that can be summed up in three values: tolerance, trading spirit and inclusiveness . . . it has never been a city of kings and bishops but it has always been a city of merchants. You find no palaces here but canal houses that date back to the days since we have been used to self-governing. And we have never forgotten to look back to see who stayed behind and should also be included.

This combination of a merchant focus on trade as well as a focus on inclusive collaboration, which allegedly characterises Amsterdam's past, can also be found in many of documents that discursively position Amsterdam as the smart city of tomorrow.

The focus of experimentation is also a pronounced feature of the smart city rhetoric in Amsterdam. Experimental initiatives and sites are discursively positioned as best practices that should be up-scaled and, in this way, they should provide the springboard to "achieve" smart urbanism and a better city for all (there are no losers in this narrative). The motto of the Amsterdam Smart City (ASC) collaborative platform provides a striking illustration of this type of reasoning:

> [we make] it possible to test new initiatives. The most effective initiatives can then be implemented on a larger scale . . . The ultimate goal of all activities is to contribute positively towards achieving CO_2 emission targets, as well as aiding the economic development of the Amsterdam Metropolitan Area. In doing so, the quality of life will improve for everyone.
>
> (ASC, 2016)

This points to another striking feature, namely that smart city experimentation is tied to an array of societal goals − in this case not only economic development but environmental sustainability as well. Of the set of 90 or so experiments aggregated under the ASC platform around 30 are explicitly framed in terms reminiscent of environmental sustainability (i.e. framings of ICTs enabling energy savings or empowering electric mobility).[6] This makes the distinction between the urban label of "smart" and other urban labels such as the "eco", "resilient" or the "circular" less pronounced. A very clear example of the mixing of urban labels has occurred in Buiksloterham. This neighbourhood was incorporated in the ASC platform website as a smart city living lab, but the individuals and organisations conducting the experiments don't use the term smart, but talk about their ambitions in terms of sustainability and the "circular economy".

5. Smart city experimentation as institutional reconfiguration

The smart city points to prospects to "manage" in new ways the flows of vehicles and people, of energy and materials. This begs the question of who will be charged with conducting this kind of management and who will optimise flows in accordance to which politically chosen objectives and ideals. From an institutional perspective, smart city experimentation is about reconfiguring the institutions that constitute urban governance, and in particular in terms of relations between the municipality, research institutes, citizens and the corporate world (Bolivar and Meijer, 2015). The relationship between the material experiments and immaterial institutions are a two-way street: experiments may fit poorly with existing institutional configurations and may thrive if institutions are reconfigured; and these re-configured institutions may gain a more secure foothold if the experiments are perceived to be successful. A clear reflection of this institutional reconfiguration is the constitution of the multiscalar actor network that supports the portfolio of smart city experiments, and the way roles are divided and coordination among actors takes shape.

Some of the canonical smart cities from scratch reflect a top-down mode of governance where the workings of the city are guided by a single "urban operation system" or where critical infrastructure is owned and operated by big technology companies. While city governments and corporations are also among the most prominent actors propagating smart city experimentation in the actually existing smart cities of Hamburg and Amsterdam, the outcome in terms of governance and power relations is far from clear.[7]

In Amsterdam, the smart city program takes the shape of metropolitan-level platform organised as Public Private Partnership. Government agencies do not play a steering role, but corporate actors and government agencies together play a facilitating role. According to Amsterdam's "chief technology officer" the smart city will especially empower the individual freedom of citizens and consumers: "the Amsterdam Smart City programme enables citizens to use the information that is available to make the choices they want to make. This means that implementation is sometimes a bit slower than with a top-down strategy" (Baron, 2013). Yet, this rhetorical focus on the citizen is not be reflected in the actual smart city experiments conducted in Amsterdam: a recent report on the ASC and its portfolio of experiments showed that citizens do not play an active partnership role in most experiments that were studied (Van Winden et al., 2016).

In Hamburg, the smart city program is directly under municipal control. Recently the city cut a deal with Cisco has been cut regarding experimentation.

This privileged cooperation has been justified as clearly confined to what the elected municipality defines as in the public interest. Moreover, after the big PR event announcing the partnership (that would be open to the inclusion of other corporate partners, it was emphasised), no follow-up information has been disseminated on the official website ever since (Meinecke, 2016). Officials in charge of the cities' "digitalisation strategy", which is nowadays highlighted as the most important smart city development in Hamburg, emphasise the mundane character of their current work in coordinating various municipal departments, and rather downplay the collaboration with CISCO as one initiative among many others. This stands in stark contrast rhetoric about the role of the citizen in Amsterdam.

6. Conclusion

We have mobilised insights from the field of sustainability transitions and aimed to provide a fresh analytical perspective on smart city developments. At the heart of this perspective are the ideas of the city as a complex patchwork of socio-technical systems, the struggle between obdurate stability and transformative change and the notion of experimentation. We have also argued that an analytical division along the lines of (1) materiality, (2) discourses and (3) institutions provides a fruitful starting point for analysis of smart city experimentation.

It should be noted that many elements of our emerging conceptualisation are still up for debate, two of which we want to highlight here. First, since the transitions literature highlights the struggle between stability and change it brings to the fore the question of whether the smart city and associated attempts to smart city experimentation should be seen as "more of the same" or as a "true agent of transformation". Within the transitions field this dichotomy is usually presented as a given whereby certain socio-technical configurations represent unsustainability and stability (e.g. the "regimes" of fossil energy, automobiles and industrial food) while other ones represent sustainability and change (e.g. the "niches" of solar power, bicycles or organic food). But for the case of smart cities there might not be a straightforward answer here. According to some, the smart city rhetoric and experimentation is all about increasing efficiency (incremental improvement) and with managing urban space through close surveillance of citizens (top-down government), which reflects incumbent systems of economic production and political governance reminiscent of stability. But according to others it represents a set of radically new technological systems through the ubiquitous integration of ICTs giving rise to a new urban experience and a new division of roles between governments, citizens and business and radically different relations and forms of governance, which reflects transformation – for better or worse. It might be the case that some of the rhetoric articulates a profoundly transformed future built environment and a prominent role for different actors than the ones that set the agenda today (discourse), but the set of actual experiments in the built environment that are supposed to give substance to this impending future (materiality) and the governance arrangements that support them (institutions) reflects a future world that is not fundamentally different in its composition. It might also be the case that some experimentation is not guided by an articulated vision of a very different future society (discourse) and supported by a coalition of actors that reflect incumbent role patterns and power relations (intuitions), but that it nonetheless becomes a stepping stones for future experimentation that leads to a surprisingly different future where relationships between people and ICT devices (materiality) or between citizens, corporations and governments (institutions) are profoundly reshuffled. Rather than a priori assuming one of the two positions, it might be more productive to analyse individual smart city experiments or collections of experiments and living labs to assess in which different ways these efforts in practice imply business as usual or a profound transformation.

Second, and related, earlier contributions to the transitions field were developed for the analysis of transitions in a single sectoral socio-technical system, such as energy, mobility or food provision. In our view, cities are best conceptualised as patchworks of socio-technical systems and their governance means that urban discourses, institutions and infrastructures are connected across the physical and/or administrative boundaries of a city. Most studies about transitions stress the importance of national-level systems and niche-, regime- and landscape-level dynamics on various time scales, but they lack a nuanced understanding of the dynamics of geographical scale, which is a prerequisite for an understanding of the urban (Hodson and Marvin, 2010; Bulkeley et al., 2011; Rohracher and Späth, 2014). To make sense of experimentation with smart urbanism, spatial scale requires more prominent attention because it implies multiple levels and sites of governance as well as multiple socio-technical systems. These multiple socio-technical systems, with rivalling or reinforcing dynamics between them, interact in complex ways and give rise to what we could call a "meta-regime" of urban planning trying to govern these dynamics. Interactions between socio-technical systems have so far received little attention in the field of sustainability transitions studies. While some notable examples exist (Konrad et al., 2008; Sutherland et al., 2015; Raven, 2007; Raven and Verbong, 2007), these studies do not take into account the ways that these interactions become locally entangled.

Finally, we should note the case studies of Amsterdam and Hamburg mobilised in this paper serve merely as an illustration of our conceptual ponderings based on the sustainability transitions literature. More thorough empirical work that is richer in detail and that is conducted in a variety of other cities and living labs will show how useful our conceptual categories and analytical framework are in describing on-going developments in experimentation with smart urbanism. More generally, the three-dimensional framework developed in this paper is essentially not limited to be applied to smart city experimentation, but can find wider application in the analysis of urban experiments and living labs that are geared to bring about very different (non-smart city) futures. In fact, our framework may serve as a useful starting point for future research in this area, for example by systematically comparing a set of experiments or living labs from a discursive, material and institutional perspective, and the ways in which a range of actors experiment with and within these structural conditions in various ways.

Notes

1 Carvalho 2014 uses the approach of Strategic Niche Management – which is part of the repertoire of the sustainability transitions field – to investigate the canonical greenfield city experiments cases of Plan IT in Portugal and Songdo in South Korea (whereas this paper is more concerned with the mundane "actually existing" smart city). The work of Luque Ayala and Marvin (2015) and Marvin et al. (2016) is more comprehensive and inspired by a wide range of smart urbanism experiences from across the globe and it is overtly critical, pointing to the political implications (whereas this paper aims to provide an analytical framework rather than a normative stance).

2 There is an emerging literature specifically on "urban sustainability transitions" (Bulkeley et al., 2011; Loorbach et al., 2016; Nevens et al., 2013; Rutherford and Coutard, 2014). Here too experimentation takes center stage, but a different set of specialised terms is used. However, the initial transitions literature talked about niche experiments, bounded socio-technical experiments, transition experiments and grassroots experiments, the later geography and urban studies literature talks about urban climate change experiments and urban living labs (see Sengers et al., forthcoming, for an overview).

3 This division according to these three dimensions is inspired by the literature on policy studies and comparative politics (Campbell, 1998; Hay, 2004; Poteete, 2003; Schmidt, 2001; Scott, 2008).

4 This division is made by combining insights from political economy (Levy and Newell, 2002) and earlier work on sustainability transitions (Avelino and Rotmans, 2009; Kern, 2011; Grin, 2010).

5 Many ambitious greenfield projects in China come with different labels attached but they embody similar ambitions and practices as other smart cities. For instance, the Tianjin-Binhai "eco-city" or Shenzhen-Pingdi "low carbon city" are not labelled as "smart city" but they include a focus on ICT-oriented innovation and impressive feats of digital monitoring reminiscent of other smart city projects.

6 Contrasts exist between Amsterdam and other Dutch cities in terms smart city vision and the focus of their experiments. The smart city roadmap of Delft has an even more pronounced focus on participation and design as starting point (Delft Smart City 2015). On the other hand, the smart city vision documents of Assen put sensors and new technology up front (Assen Sensor City 2016) – this attests to very different coalition of actors claiming the term for a particular city.

7 Especially the recent book by Anthony Townsend, "Smart Cities: Big Data, Civic Hackers, and the Quest for a New Utopia" presents the bottom-up versus top-down interpretation of the smart city as two opposing camps "that are bound to come to blows", the outcome of which would then determine what kind of actually-existing-smart-city we will get (Townsend, 2013).

References

ASC. (2016). Amsterdam Smart City platform. Retrieved from https://amsterdamsmartcity.com/ (accessed 20 July 2016).

Assen Sensor City. (2016). Assen Sensor City: Innovatieplatform voor livinglab Assen. Retrieved from www.sensorcity.nl/ (accessed 20 July 2016).

Avelino, F. and Rotmans, J. (2009). 'Power in transition: An interdisciplinary framework to study power in relation to structural change'. *European Journal of Social Theory*, *12*(4), 543–569.

Baron, G. (2013). 'Smartness from the bottom-up: A few insights into the Amsterdam Smart City programme'. *Metering International, 3,* 98–101

Bijker, W., Hughes, T. and Pinch, T. (1987). 'The social construction of technological systems: New directions in the sociology and history of technology'. Cambridge, MA: MIT Press

Bolivar, M. and Meijer, A. (2015). 'Smart governance: using a literature review and empirical analysis to build a research model'. *Social Science Computer Review.* In Press.

Bulkeley, H., Breitfuss, M., Coenen, L., Frantzeskaki, N., Fuenfschilling, L., Markus, G., . . . Voytenko, Y. (2015). Theoratical framework. Working paper on urban living labs and urban sustainability transitions: GUST.

Bulkeley, H. and Castán Broto, V. (2013). 'Government by experiment: Global cities and the governing of climate change'. *Transactions of the Institute of British Geographers, 38,* 361–375.

Bulkeley, H., Castán Broto, V., Hodson, M. and Marvin, S. (2011). Cities and low carbon transitions. Routledge: London

Campbell, J. (1998). 'Institutional analysis and the role of ideas in political economy'. *Theory and Society, 27,* 377–409.

Campbell, T. (2012). *Beyond Smart Cities: How cities network, learn, and innovate.* Abingdon, UK: Earthscan.

Carvalho, L. (2014). 'Smart cities from scratch? A socio-technical perspective'. *Cambridge Journal of Regions, Economy and Society, 8*(1), 43–60.

Deakin, M. and Al Waer, H. (2012). From intelligent to smart cities.London: Routledge.

De Jong, M., Joss, S., Schraven, D., Zhan, C. and Weijnen, M. (2015). 'Sustainable – smart – resilient – low carbon – eco – knowledge cities: Making sense of a multitude of concepts promoting sustainable urbanization'. *Journal of Cleaner Production, 109,* 25–38

Delft Smart City. (2015). Delft Smart City. Retrieved from www.delft.nl/Bedrijven/Stad_van_innovatie/Delft_Smart_City (accessed 10 May 2016)

Evans, J., Karvonen, A. and Raven, R. (2016). The experimental city. London: Routledge.

Geels, F. (2014). 'Regime resistance against low-carbon transitions: Introducing politics and power into the multi-level perspective'. *Theory, Culture and Society, 31*(5), 21–40.

Greenfield, A. (2013). *Against the smart city.* New York: Do Projects.

Grin, J., Rotmans, J., Schot, J., Geels, F. and Loorbach, D. (2010). *Transitions to sustainable development: New directions in the study of long term transformative change.* New York: Routledge.

Hajer, M. (2006). 'Doing discourse analysis: Coalitions, practices, meaning'. In M. v. d. Brink and T. Metze (Eds.), *Words matter in policy and planning. Discourse theory and method in social science* (pp. 65–76). Netherlands Geographical Studies.

Hajer, M. and Dassen, T. (2014). *Smart about cities.* The Hague: PBL

Hajer, M. A. (1995). *The politics of environmental discourse: Ecological modernization and the policy process.* New York: Oxford University Press.

Hay, C. (2004). 'Ideas, interests and institutions in the comparative political economy of great transformations'. *Review of International Political Economy, 11,* 204–226.

Hodson, M. and Marvin, S. (2010). 'Can cities shape socio-technical transitions and how would we know if they were?' *Research Policy, 39,* 477–485.

Hodson, M. and Marvin, S. (2014). *After sustainable cities.* New York: Routledge.

Hollands, R. (2008). 'Will the real smart city please stand up?' *City, 12,* 303–320.

Hommels, A. (2005). 'Studying obduracy in the city: Towards a productive fusion between technology studies and urban studies'. *Science Technology and Human Values, 30*(3), 323–351.

Kern, F. (2011). 'Ideas, institutions and interests: Explaining policy divergence in fostering 'system innovations' towards sustainability'. *Environment and Planning C, 29*(6), 1116–1134.

Kitchin. R. (2014). 'Making sense of smart cities: Addressing present shortcomings'. *Cambridge Journal of Regions, Economy and Society, 1*(6), 131–136.

Konrad, K., Truffer, B. and Vosz, J.P. (2008). 'Multi-regime dynamics in the analysis of sectoral transformation potentials. Evidence from German utility sectors'. *Journal of Cleaner Production, 16*, 1190–1202.

Latour, B. (1996). *Aramis, or the love of technology.* Cambridge, MA: Harvard University Press.

Levy, D. and Newell, P. (2002). 'Business strategy and international environmental governance: Toward a neo-Gramscian synthesis'. *Global Environmental Politics, 2*(4), 84–101.

Lindblom, C. (2001). *The market system. What it is, how it works, and what to make of it.* New Haven, CT: Yale University Press.

Loorbach, D., Wittmayer, J., Shiroyama, H., Fujino, J. and Mizuguchi, S. (2016). *Governance of urban sustainability transitions: European and Asian experiences.* Toyko: Springer Japan.

Luque-Ayala, A. and Marvin, S. (2015). 'Developing a critical understanding of smart urbanism?' *Urban Studies, 52*, 2105–2116.

Markard, J., Raven, R. and Truffer, B. (2012). 'Sustainability transitions: an emerging field of research and its prospects'. *Research Policy, 41*, 955–967.

Marvin, S., Luque-Ayala, A. and McFarlane. C. (2016). *Smart urbanism: Utopian vision or false dawn?* New York: Routledge.

Meadowcroft, J. (2011). 'Engaging with the politics of sustainability transitions'. *Environmental Innovation and Societal Transitions, 1*(1), 70–75.

Meinecke, S. (2016). Smart cities und die Mobilität der Zukunft. Hamburg Municipality website. Retrieved from www.hamburg.de/smart-city/ (accessed 1 July 2016).

Nevens, F., Frantzeskaki, N., Loorbach, D. and Gorissen, L. (2013). 'Urban transition labs: Co-creating transformative action for sustainable cities'. *Journal of Cleaner Production, 50*, 111–122.

Poteete, A. (2003). 'Ideas, interests and institutions: Challenging the property rights paradigm in Botswana'. *Governance, 16*, 527–557.

Raven, R. (2007). 'Co-evolution of waste and electricity regimes: Multi-regime dynamics in the Netherlands (1969–2003)'. *Energy Policy, 35*(4), 2197–2208

Raven, R. and Verbong, G. (2007). 'Multi-regime interactions in the Dutch energy sector. The Case of Combined Heat and Power in the Netherlands 1970–2000'. *Technology Analysis and Strategic Management, 19*(4), 491–507.

Rohracher, H. and Späth, P. (2014). 'The interplay of urban energy policy and socio-technical transitions: The eco-cities of Graz and Freiburg in retrospect'. *Urban Studies, 51*, 1415–1431.

Rutherford, J. and Coutard, O. (2014). 'Urban energy transitions: Places, processes and politics of socio-technical change'. *Urban Studies, 51*(7), 1353–1377.

Schmidt, V. (2001). 'The politics of economic adjustment in France and Britain: When does discourse matter?' *Journal of European Public Policy, 8*, 247–264.

Scott, W.R. (1995). *Institutions and Organizations.* Thousand Oaks, CA: Sage.

Scott, W.R. (2008). *Institutions and Organizations: Ideas and interests.* London: Sage.

Sengers, F., Wieczorek, A. and Raven, R. (2016). 'Experimenting for sustainability transitions: A systematic literature review'. *Technological Forecasting and Social Change.* In Press.

Shelton, T., Zook, M. and Wiig, A. (2015). 'The "actually existing smart city" '. *Cambridge Journal of Regions, Economy and Society, 8*, 13–25.

Smink, M., Hekkert, M. and Negro, S. (2015). 'Keeping sustainable innovation on a leash? Exploring incumbents' institutional strategies'. *Business, Strategy and the Environment, 24*, 86–101.

Smith, A. and Raven, R. (2012). 'What is protective space? Reconsidering niches in transitions to sustainability'. *Research Policy, 41*, 1025–1036.

Söderström, O., Paasche, T. and Klauser, F. (2014). 'Smart cities as corporate storytelling'. *City, 18*(3), 307–320.

Stadt Hamburg. (2016). *Hamburg – SmartCity – City for innovation*. Hamburg.

Stadt München. (2016). München als Smart City. Retrieved from www.muenchen.de/rathaus/Stadtverwaltung/Referat-fuer-Stadtplanung-und-Bauordnung/Stadtentwicklung/Perspektive-Muenchen/Smart-City.html (accessed 23 August 2016).

Stegmaier, P., Kuhlmann, S. and Visser, V. (2014). 'The discontinuation of socio-technical systems as a governance problem'. In S. Borrás and J. Edler (Eds.), *The governance of socio-technical systems: Explaining change* (pp. 111–131). Cheltenham, UK: Elgar.

Sutherland, L., Peter, S. and Zagata, L. (2015). 'Conceptualising multi-regime interactions: The role of the agriculture sector in renewable energy transitions'. *Research Policy, 44*, 1543–1554.

Townsend, A. (2013). *Smart cities: Big data, civic hackers, and the quest for a new Utopia*. London: Norton.

Turnheim, B. and Geels, F. (2013). 'The destabilisation of existing regimes: Confronting a multi-dimensional framework with a case study of the British coal industry (1913–1967)'. *Research Policy, 42*(10), 1749–1767.

Van den Bergh, J. (2012). 'EIST one year: Something to celebrate?' *Environmental Innovation and Societal Transitions, 4*, 1–6.

Vanolo, A. (2014). 'Smartmentality: The smart city as dis-ciplinary strategy'. *Urban Studies, 51*(5), 883–898.

Van Winden, W., Oskam, I., Van den Buuse, D., Schrama, W. and Van Dijck, E. (2016). *Organising smart city projects: Lessons from Amsterdam*. Amsterdam: Hogeschool van Amsterdam.

Voytenko, Y., McCormick, K., Evans, J. and Schliwa, G. (2016). 'Urban living labs for sustainability and low carbon cities in Europe: Towards a research agenda'. *Journal of Cleaner Production, 123*: 45–54.

PART II

Practices of ULL

6

INTERMEDIATION AND LEARNING IN STELLENBOSCH'S URBAN LIVING LAB

Megan Davies and Mark Swilling

1. Introduction

Stellenbosch, a major university town in the Western Cape of South Africa, aims to position itself as a leading and innovative African city-region. This is captured in the vision for Stellenbosch Municipality (SM) to be the "Innovation Capital of South Africa". Despite being a relatively small municipality, it is faced with myriad developmental and urban sustainability challenges, experienced predominantly in the town of Stellenbosch, the municipality's largest urban node. These include substantial infrastructure backlogs due to long-term under-funding and insufficient provision for future demand, and ad hoc spatial development which both entrenches and exacerbates spatial exclusion and economic disparity. This complex, and seemingly intractable, dynamic hinders the municipality's mandate of delivering sufficient and equitable basic services and enabling inclusive local economic development. The realisation of an ecologically, socially and economically sustainable development trajectory for the region is hindered by a lack of internal strategic coordination and long-term integrated planning. Historically, this is evident in the absence of a coherent development strategy coupled with a lack of coordination with private sector, civil society and research institutions, particularly with Stellenbosch University (SU), an internationally renowned research institution around which the local economy is anchored.

It is within this context that a set of relationships opened up over the last decade and culminated in a unique governance arrangement between the university and municipality, underpinned by a commitment to transdisciplinary research. The Rector-Executive Mayor Forum (REMF), set up in 2005, resulted in in the establishment of two sub-committees in 2014 – the Integrated Planning Committee (IIP) and the Infrastructure Innovation Committee (IIC). Constituting municipal officials and political representatives, selected private sector players and university researchers and administrators, the IIC and IPC represented the coming together of a diverse array of stakeholders with distinctive objectives and visions for the future, in an effort jointly to tackle the region's development and sustainability challenges. The progress thereof is demonstrated in two major outputs from the REMF, these include the draft Stellenbosch Spatial Development Framework (SSDF) produced by the IPC in 2015 and the Stellenbosch Quo Vadis document produced by the IIC in 2014. Intended as an input to the overarching Spatial Development Framework (SDF) for the municipal area, the Stellenbosch town plan as it was informally referred to, comprised a formal planning process overseen by the Planning and Economic Development department at SM. The SSDF was done in collaboration with the Sustainability Institute (SI) as appointed project management, under the banner of the IPC and the various stakeholders represented in the forum. Unlike the SSDF process that was required to comply with formal procedures within the municipality, the Quo Vadis Document had the advantage of being an internal, informal guiding document that in turn provided cohesion around strategic direction for the IIC. Together, these documents reflect the outcome of nearly five years of debate and engagement between the institutions, particularly around issues of sustainable infrastructure development and spatial planning. They provide the basis for a potential large-scale program of collaborative innovation and experimentation in Stellenbosch over the coming years. As important as the strategic content of these documents, is the process that led to their production. Between 2011 and 2015, the initiative resulted in unprecedented cooperation between different departments within SM, and encouraged more meaningful integrated planning between SM and SU.

As an emergent outcome of this dynamic collaboration, Stellenbosch's urban living laboratory (ULL) (Evans, Karvonen and Raven, 2016; Voytenko, McCormick, Evans and Schliwa, 2015) has its foundation in the commitment from key SU researchers from the School of Public Leadership (SPL) and the Centre for Complex Systems in Transition (CST) to developing transdisciplinary research capabilities at the institution (Davies, 2016; van Breda and Swilling, forthcoming). Overall, the REMF demonstrates a strategic reorientation of the university towards more meaningful local community interaction and social impact; an intellectual commitment to transdisciplinary research within certain sections of the research institution, and a receptiveness by the municipality to collaboration and partnership to address shared challenges of socio-economic development in the region given their financial and human-capacity limitations. In this way, the various motives of the university researchers and administrators, and municipal officials and politicians are recognised. Between 2011 and 2015, the REMF as Stellenbosch's ULL responded to these various interests in such a way that the forum and its sub-committees functioned effectively, and thus realised the draft SSDF and internal Quo Vadis Document as concrete outputs.

As an investigation of the particularities of how this ULL has been designed and facilitated, the first section of this paper explores its roots in sustainability science and transdisciplinary research at SU. Of particular importance is the cultivation of a space of intermediation and learning that shaped the REMF as a hybrid space or "learning agora", conducive to innovation and experimentation. This paper is, therefore, a contribution to the growing literature on "urban experimentation" (Evans et al., 2016). Thereafter, the second section will unpack the evolution of the Stellenbosch ULL in terms of how the REMF, IIC and IPC unfolded between November 2013 and April 2015. The chapter will conclude with a reflection of the unique design and evolution of the ULL.

2. Foundations of collaborative governance in Stellenbosch

2.1. Stellenbosch Municipality: An overview

Stellenbosch Municipality is situated roughly 50 kilometres from Cape Town in the Western Cape Province of South Africa. The municipal region covers around 900 km² and has a total population of roughly 155,000 people according to municipal figures (Stellenbosch Municipality, 2016). This figure is contested, with other sources indicating a population as high as 220,000 (Swilling, Sebitosi and Loots, 2012). Regardless, the region, and the town of Stellenbosch in particular, is a focal point for growth.

The municipality is located in the heart of the Cape Winelands that is dominated by agricultural land of high historic, cultural and aesthetic value and globally important natural habitats (Stellenbosch Municipality, 2016). The area has a long history with the town of Stellenbosch having been established in the late 1600s. It now includes well developed tourism, education, research and agricultural industries while moving towards a more tertiary services-oriented economic focus supported by a growing manufacturing and construction industry. Set within this network of agricultural and conservation areas is a network of urban settlements. The municipality has two major towns – Stellenbosch and Franschhoek – and a range of other formal, more rural and informal settlements (Stellenbosch Municipality, 2016). Stellenbosch is the largest of the municipality's 14 official urban nodes and is the urban centre around which the local economy is anchored (IIC, 2014). The town is one of South Africa's oldest formal settlements and like most South African communities, it exhibits considerable inequality (Ewert, 2013). It also has to contend with many of the same challenges (Nicks, 2012). The town's urban fabric is reminiscent of the "suburban dream of apartheid planning given its fragmentation and physical segregation", even so, Stellenbosch "offers an urban experience of a quality and intensity unique among South African towns" (Nicks, 2012, p. 24).

The many faces of Stellenbosch deliver vastly incongruent experiences for residents, employees and visitors. And despite Stellenbosch's apparent prosperity, its flourishing tourism, manufacturing, financial and agricultural sectors, "this picturesque town has its fair share of 'ugly' poverty" (Ewert, 2013, p. 1; Nicks, 2012). This highly divided town thus reflects a microcosm of the wider patterns of inequality within South Africa, and indeed, those institutionalised in global patterns of disproportionate production and consumption (Swilling and Annecke, 2012). Despite this seemingly dismal set of circumstances, Stellenbosch is also endowed with "extraordinary intellectual capacity with the university at the heart of the community, social diversity, financial resources, creative potential, high value eco-systems, spiritual energy and some of South Africa's most vibrant grassroots social movements in its poorest areas" (Nicks, 2012, p. 31). It is widely recognised that Stellenbosch and the greater Stellenbosch region is characterised by a remarkable and unique concentration of capabilities, resources and opportunities, a favourable position from which to contend with these development challenges (Swilling et al., 2012).

2.2. Initiating institutional cooperation

The institutional conditions enabling partnership and collaboration between the municipality and the university have been made possible, primarily, through the establishment of the REMF in 2005 (Swilling, 2014). In the early 2000s, the Rector and Vice Chancellor Professor Chris Brink at the time introduced the vision of Stellenbosch as a university town and undertook to reposition SU within a network or league of internationally renowned university towns (Swilling, 2014). The implication of this was the realisation that in order to frame Stellenbosch as a university town, the university needed to rejuvenate its local roots, leverage its connections to a particular context and embed itself further in the dynamics of a distinctive municipal and regional system. What became apparent, however, was that the relationship between the two institutions was limited and in an effort to facilitate greater coordination between the university and municipality on a very practical level and to address greater strategic alignment, monthly meetings were initiated involving the Rector, Executive Mayor and key officials from the respective administrations, plus academics.

The sustained effort to build this partnership between SU and SM further served to reinforce the university's positioning within a global discourse – one which has been hugely successful and is now widely accepted. Distinctive from its identity as a research institution, cooperation with the municipality was equally advantageous for the university's management and administration. Similarly, for the municipality, a partnership with the university presented both pragmatic and strategic opportunities – as a possible avenue through which to address severe financial and human capacity deficits for example, through the application of research resources towards matters of joint importance to the respective administrative bodies, (traffic congestion, a lack of affordable housing among others) as well as to align with cutting-edge innovations and technologies made available through the cooperation with the university as an internationally renowned research institution (Davies, 2016).

The introduction of The Hope Project, a long-term strategic plan for the university launched in 2010 and pioneered by the previous vice Chancellor and

Rector Professor Russell Botman, with its strategic focus on "science for society" and cultivating a "pedagogy of hope", further entrenched the positioning of the university as an institution embedded in its local context, and one cognisant of its responsibility to wider regional, national and international research imperatives (SU, 2010, 2012). However, the legitimation of the REMF collaboration under the leadership of Professor Chris Brink was in terms of the positioning of SU in a global discourse, the orientation under Professor Russel Botman was towards achieving the Millennium Development Goals (MDGs) (Swilling, 2014). This attitude towards collaboration has been carried forward by university leadership as "the university endeavours to create the conditions that will ignite the imagination of scientists to solve problems in creative ways through basic and applied research and through multi-, inter- and transdisciplinary academic activities" (Swilling et al., 2012, p. 4). In the words of Prof Russel Botman, the Rector at the time,

> the university has a social contract with the town and all of its people. Unlike the conventional use of the term, which seeks to provide a legitimate basis for political authority, the university's pact with Stellenbosch entails a willingness to be of service to the community.
>
> (Swilling et al., 2012, p. xvii)

Currently, the university is under the leadership of Professor Wim de Villiers, and while the strategic vision might have evolved, the three strategic pillars of the institution capture the broad areas of involvement of SU: teaching and learning, research and community interaction (SU, 2010, 2012).

From the onset, the REMF provided a forum where officials of the university and municipality met at least once a month to discuss issues of mutual concern and to coordinate their efforts in the promotion of human development (Swilling et al., 2012). From the municipality's perspective, a partnership with the university and more specifically the CST and the SI was advantageous as it would support developing innovations for building of a green economy in the municipality (Swilling et al., 2012; Swilling, 2014). As Executive Mayor at the time, Conrad Sidego wrote in *Sustainable Stellenbosch*:

> Innovations, however, do not happen just because they are needed. World-wide experience shows that spaces for engagement, dialogue, exploration and creativity need to be opened up and fostered, because it is from these kinds of spaces that innovations tend to emerge. Innovations are usually the outcome of intense interactions between researchers, investors and practitioners who manage to build sufficient trust so that they can jointly tackle complex problems. Without trust and these spaces for innovation, we will not overcome the challenges faced by Stellenbosch.
>
> (Swilling et al., 2012, p. xi)

The municipality's focus on innovation is demonstrated in its vision to become "The Innovation Capital of South Africa" and corresponding strategic priorities namely: preferred investment destination, greenest municipality, safest valley, dignified living and good governance and compliance (Stellenbosch Municipality, 2016).

2.3. Sustainability science, transdisciplinary research and the "learning agora"

The recent emergence of the distinctive domain of sustainability science is indicative of the shifting demands on knowledge production and responds to the complexity of sustainability challenges (Burns, Audouin and Weaver, 2006; Hadorn, Bradley, Pohl, Rist and Wiesmann, 2006; Jahn, 2008; Kajikawa, Tacoa and Yamaguchi, 2014; Kauffman and Arico, 2014; Regeer and Bunders, 2009). In many ways, this is reflected in the strategic reorientation of SU as described above. The distinctive mandate for sustainability science is developing knowledge that is "user-inspired and, at its best, provides solutions to real-world problems encountered for the needs of a sustainability transition" (Kates, 2010 in Kauffman and Arico, 2014, p. 413). Kajikawa et al. (2014, p. 432) emphasise how sustainability science must take place in the real-world and so "we have no alternative but to engage society in collaboration and to attempt change in an environment that requires trans-disciplinary practices".

Regarding the setting for transdisciplinary approaches, these "require new rules and norms that merge both academic and practice-based requirements and mandates" (Polk, 2014, p. 449). Polk (2014) suggests that transdisciplinary approaches must create a space where science and policy can meet and interact on equal terms. Polk (2014) suggests that this hybrid space must exist alongside, but not entirely separate from, the formal confines of disciplinary, administrative and political cultures. Instead, these meeting places need to be highly embedded within both spheres in order to enable actors to break the boundaries between diverse knowledge and expertise (Polk, 2014). These sites of interaction are critical for producing the necessary participation and knowledge integration that can more effectively bridge the gap between science and policy spheres (Polk, 2014).

Within these institutional spaces of intermediation, transdisciplinarity is about joint problem solving and mutual learning as part of a social learning process (Reyers et al., 2010; Scholz, Mieg and Oswald, 2000; Schneider and Rist, 2014 van Breda and Swilling, forthcoming). The generation of the three types of transdisciplinary knowledge (systems, target and transformation knowledge) "takes place within an interactive learning process, involving discussion and negotiation, and leading to a common knowledge base which may fulfil scientific standards (validity), demands of the political and administrative systems (policy relevance) but also social robustness (societal relevance)" (Schauppenlehner-Kloyber and Penker, 2015, p. 59).

The development of transdisciplinary research at SU is detailed in van Breda and Swilling's forthcoming 'The guiding logics and principles for designing emergent transdisciplinary research processes: learning experiences and reflections from a South African transdisciplinary case study: 2011–2016'. While a broad commitment from researchers to transdisciplinary research was fundamental right from the onset of the REMF, the conception of "emergent transdisciplinary design" is a recent and novel articulation of a particular logic of transdisciplinary research (van Breda and Swilling, forthcoming). As such "a core element of this approach is that the research process is designed as it unfolds; that is, it transforms as it emerges *from* and *within* the fluid context" (van Breda and Swilling, forthcoming). The REMF was one context in which the tenets of emergent transdisciplinary design were surfaced and tested, as part of researchers' critical reflection on "science with society" at SU.

Emergent transdisciplinary design is a response to doing research in a developing world context where the goal is "to develop solution-oriented or transformative TDR approaches capable of not only explaining and understanding the complex societal challenges currently being faced in the world, but also of changing or transforming these challenges" (van Breda and Swilling, forthcoming).

For van Breda and Swilling (forthcoming), engagements within an emergent transdisciplinary design process allows for deep immersion into particular socio-historical settings. It also allows for the emergence of social learning experiences of researchers, co-researchers and individual social actors gathered as they learn to navigate the many theoretical and practical challenges facing them, with the result of dynamically shaping the development of the approach itself. It also creates the space for experimenting with radically different approaches, critical reflection and theorising.

Such an interactive and holistic learning process is about the personal development of stakeholders in the form of the reflexivity, questioning and possible integration of underlying assumptions, knowledge, goals and values (van Breda and Swilling, forthcoming; Wittmayer and Schäpke, 2014). Multi-loop transformative learning is one of the five core guiding principles of emergent transdisciplinary design where "embedded, experiential, learning-by-doing" allows for the co-production of transformative knowledge and the identification of strategic interventions that bring about incremental social change (van Breda and Swilling, forthcoming). Additional principles include, perturbing the system, exaptation and bricolage, allowing for emergence and absorbing complexity (van Breda and Swilling, forthcoming). Seen together these principles encourage researchers to embrace the provisional nature of the research context and research design process; appreciate the goal of understanding the meaning of an unfolding situation and the process is thus adaptive, open and contextual (van Breda and Swilling, forthcoming).

Pohl et al. (2010, p. 270) offer the concept of an interactive and permeable learning "agora" with the transdisciplinary approach creating an in-between space "in which the boundaries are provisionally blurred". This resonated with notion of emergent transdisciplinary design proposed by van Breda and Swilling (forthcoming). It is within the agora that according to Wittmayer and Schapke (2013, p. 485) "science and society address real-world problems, generate knowledge, formulate solutions and pilot actions for a more sustainable future". Further the "learning agora" supported through emergent transdisciplinary design aligns with Polk's (2014) recommendation that transdisciplinary spaces of intermediation need to find a balance between being embedded in and suspended from formal structures. Pohl et al. (2010, p. 270) continue to explain how interaction within the agora contributed to learning processes: "the resulting 'messiness' of 'divided identities' is the necessary condition for engaging with 'others' and ultimately helping to reshape the involved groups' 'perceptions, behaviour and agendas that occur as a function of their interaction'". The purpose of this messiness is summarised by Wittmayer and Schapke (2013, p. 485): "overall, these spaces are characterised by the co-construction of social reality by their participants – common futures, lived reality, social identities and roles are all negotiated within them". According to van Breda and Swilling (forthcoming), these less formalised messy spaces for co-producing solutions-oriented knowledge have been under-theorised, hence their emphasis on an abductive methodology, rather than the usual inductive and deductive methodologies.

2.4. Transdisciplinary research at SU and the REMF as a space of intermediation

Polk's (2014) notion of a hybrid space, and Pohl et al.'s (2010) learning agora, in which stakeholders from science, policy and society can meet and interact on (ideally) mutual terms resonates strongly with the notion of emergent transdisciplinary design that best characterises what emerged within the REMF (van Breda and Swilling, forhtcoming). Van Breda and Swilling (forthcoming) emphasise that "the particular challenge for doing transdisciplinary research is the engagement with non-academic actors to enable joint problem formulation, analysis and transformation, because how this is being done differs fundamentally from context to context". Thus, what distinguishes the approach to transdisciplinary research at SU is the recognition that "transdisciplinary research in a developing world context cannot be done in exactly the same way as in a developed world context, because the material and social conditions assumed by the latter are simply not present in the former" (van Breda and Swilling, forthcoming). In essence, what has been pioneered is a solutions-oriented approach to transdisciplinary research in developing world contexts that are characterised by high levels of social fluidity, where the research process is designed as it unfolds and where the goal is to co-generate knowledge applicable to context-specific, complex social challenges (van Breda and Swilling, forthcoming.). And furthermore, in the context of the Anthropocene, the fundamental purpose of producing knowledge "must to be to contribute to social change in the direction of producing more just and sustainable societal outcomes" (van Breda and Swilling, forthcoming).

The REMF evolved into a formalised experiment in trans-disciplinary research and a unique governance arrangement between these two institutions. At no point was the REMF framed as a research endeavour driven and owned by the university and so a transdisciplinary methodology was not explicitly or neatly pursued. The initiative was operated in a manner that attempted to emulate this joint responsibility – meetings were scheduled monthly and alternated between being hosted in university venues and municipal chambers and chaired by either the Rector or the Executive Mayor depending on the venue. This shared responsibility set a precedent for a similar attitude in initiatives borne from the REMF, such as the IIC and IPC. As the REMF was recognised at the highest level of both SU and SM, this had implications (both problematic and advantageous) for the positioning and legitimacy of its sub-committees. Its recognition was heavily reliant upon the reputation, rank and credibility of the key officials that had driven the REMF over the last decade. This required discerning and pragmatic facilitation and speaks to the unique demands on transdisciplinary researchers within spaces of intermediation and learning such as this.

Drawing on the in-depth case study narrative presented by Davies (2016), reflections by participants in the REMF and its substructures during the period November 2013 to April 2015, demonstrate some of the nuances of Stellenbosch's ULL.

Recollections by forum members of the relationships between the university and the municipality preceding the establishment of the REMF are of it being a distant, mistrustful and often antagonistic one. Without any formal structures to facilitate communication and collaboration, there was no real understanding between the institutions regarding the respective strategic priorities, and according to the university's planning division, this was unfortunate since the university comprises such a large chunk of the town. This is reflected in the words of the division's director at the time, that "if the university sneezes, the town gets a cold". This so-called "town-gown" tension meant that initial discussions as part of the REMF were perceived as somewhat reluctant, reserved and stiff. However, sufficient support and leadership from both institutions meant that the initiative was pursued and as discussions about issues of mutual importance and relevance continued, more amicable relationships were developed. This emphasises the sentiment of one of the participants that the "real achievement (of the SITT and the IIC) was the creation of the relations of trust and understanding between officials and between officials and university representatives" (Davies, 2016).

Participation in these sub-structures offered stakeholders from SU and SM opportunities to interact in a learning agora that provided a measure of protection, creativity, discipline and strategic direction. The emergence of a distinctive culture within the IIC was enabled by a diversity of engagements, from formal meetings, workshops, outings, to informal get-togethers, that encouraged rich interactions open to a wide range of participants. A significant ritual that supported the cultivation of an open, critical and creative learning agora was the informal drinks that followed IIC meetings every second Friday. Postgraduate student researchers and IIC guests were invited to attend and it became common practice for those who were able to stay longer, to move from the regular meeting venue on the university campus to a venue in the town.

The informal and conversational tone of IIC meetings did not detract from it being a space of engagement where agenda-setting and long-term collective goal-setting and experimentation was possible. According to a senior municipal official,

> from an administration point of view, the big positive of the IIC is that it pull you out of the normal day to day operational issues to a much more strategic level so you are encouraged to think on a different level, more practically and holistically.
>
> (Davies, 2016)

Another municipal participant conveyed a similar sentiment saying that "the IIC lets you to step into a different space that allows you to get perspective to see how things are developing, compared to your day to day work" (Davies, 2016). As a political representative of council, another forum member said that "the IIC lets me move away from the dog-eat-dog culture in politics" (Davies, 2016). In terms of roles and identities, the IIC was experienced as a safe space for officials to flesh out ideas and learn from the collective expertise of the group. Similarly, a municipal director reflected that even with demanding workloads, participation in the IIC was sustained, indicative of the benefit that that this engagement space provided for municipal officials in particular. As such, the learning agora provided a space for honest debate where ideas could be interrogated and reshaped, supported by trusting relationships.

Beyond an intermediary body in itself, the fluid and dynamic IIC and IPC structures functioned more as intermediation spaces supportive of learning processes. Propelled and shaped by different leadership agendas and a dynamic changing environment, the REMF evolved as a collaborative platform that both institutions recognise as in some way supportive of the fulfilment of their respective mandates. Participation in these sub-structures offered stakeholders from SU and SM opportunities to interact in a space of intermediation that activated, coordinated and sustained particular combinations of resources and capabilities that would not otherwise have been possible. Municipal administrators utilised this collaboration as a way of activating pertinent resources and capabilities outside of their jurisdiction. For researchers, this space opened up novel research opportunities for real-world problems to become the focus of applied sustainability research. Thus, the facilitation of collaborative governance in SM by key researchers and students to some extent realised the ambitions of sustainability science: allowing real-world problems to become the drivers for transdisciplinary research and learning.

3. Innovation and experimentation in Stellenbosch's urban living lab

The previous section demonstrated the roots of Stellenbosch's ULL in the learning agora facilitated by the REMF collaboration. As a designated space for experimentation and innovation, the concept of the ULL aligns strongly with that of a hybrid institutional space supported by transdisciplinary research. Parallels can be drawn with the concepts of encounter spaces (Valentine, 2007) and "alternative milieu" (Longhurst, 2015) to demonstrate the various attempts, across disciplinary fields, to leverage the potential of institutional hybridity in transformation processes. The emergent transdisciplinary design of encounter spaces demonstrates the significance of the kinds of socio-spatial engagements that facilitate meaningful contact in support of mutual understanding and shifting perspectives. Leitner's (2012) work illustrates how spaces of encounters are structurally, socially and spatially mediated. Pereira, Karpouzoglou, Doshi and Frantzeskaki (2015, p. 6035) offer the concept of a "safe space" in the context of sustainability transitions as one where participants are free "to think freely without the weight of a disciplinary history or institutional commitments to a given approach that may constrain dialogue, co-create and prepare innovative ideas and interventions". Longhurst's (2015, p. 184) complementary concept of the alternative milieu in the support of sustainability innovation is another angle on the notion of a learning agora – "a localised density of countercultural institutions, networks, groups and practices that creates a particular form of geographical protection for the emergence of different forms of sustainability experiments".

With its core focus on aligning strategic planning and implementation processes between the institutions, the IPC had a strong spatial orientation and facilitated the crafting of a unique approach to drafting the SSDF. Called "Shaping Stellenbosch", this was a spatial planning initiative informed by the principles of Appreciative Inquiry (Cooperrider and Srivastva, 1987). In collaboration with the SI and the researchers from the SPL and CST, this approach demonstrated in practice how the thoughtful design of an innovative and inclusive "ideas gathering campaign" using appreciative inquiry methods made it possible for citizens, officials, students and professionals alike to contribute towards generating a positive vision for urban development within the parametres of a statutory spatial planning process. This participatory process, combining "top-down" and "bottom-up" strategic inputs, culminated in a draft SDF for the town of Stellenbosch around the vision of a *compact, inclusive and sustainable town*. Despite the fact that it has not been officially adopted as a municipal development framework due in part to cumbersome administrative procedures, much of the thinking generated through the process has had wider impacts on the decision making and strategic direction of municipal officials (Davies, 2016).

The IIC was set up at the end of 2013 as a continuation of a previous committee, namely the Strategic Infrastructure Task Team (SITT), which was shut down in 2012 because of political contestation within the muncipality's leadership structures. This new forum drew from the progress made in the SITT but embodied a tactical reorientation to account for a wider strategic focus, led by the Mayor's vision of Stellenbosch Municipality as the Innovation Capital of South Africa. The IIC, with this explicit focus on innovation, succeeded in facilitating a coherent development strategy for SM – which fed into and influenced the SSDF process – integrating finance, spatial and infrastructure planning as the foundation for achieving its ultimate purpose of putting together a Strategic

Infrastructure Plan (SIP) linked to a Financial Plan (FP). Encapsulated in the Quo Vadis document (where are we going?), an internal strategic document, this collaboration introduced a vision of a sustainable transit-oriented development approach that transcended the tension between Stellenbosch's ultra-conservative "heritage/conservation" approach and the developer-driven "sprawl" approach. Although not exclusive, this regional vision reorients SM to becoming more proactive in enabling infrastructure, finance and spatial planning to prioritise integrated public transport-oriented and infrastructure-led development. It further ensures that SM strategic planning is aligned with wider national and provincial planning imperatives guided and informed by the National Development Plan and the Integrated Urban Development Framework.

The IIC and IPC made significant progress in 2014 in particular, with the creation of the Quo Vadis Document and the draft SSDF. This document described the narrative of moving toward a new spatial vision for a *compact, inclusive and sustainable town*, and outlined a strategic framework of opportunity for development, supported by a public transport-oriented, infrastructure-led development logic. Even as the highest expression of the commitment to institutional collaboration, the REMF was extremely susceptible to wider political and institutional dynamics. The REMF and its sub-structures were continually contested however this was not directly connected to the contributions or failings of the committees themselves. Instead, as fluid and emergent processes, the committees were often used as "footballs" in the institutional and political "battles" between municipal departments, between the council and the administration and between different informal and formal factions within and between, the university and the municipality. Internal fallouts had implications for the framing of the IIC and IPC – for example, conflict between members of the Mayoral Committee served to undermine the IIC in the eyes of the wider council as they "used" the committee to further their own political or personal agendas. Prominent actors within the intermediation space were advantageously positioned to facilitate and mediate internally but in many cases were limited in how they could advocate for the committees within wider formal processes. This was where high level support was useful as the rank of prominent and powerful stakeholders within the municipality and the university could be leveraged. A significant event was the sudden death of Rector Prof Russel Botman in late 2014, which essentially closed the chapter on the REMF as it had operated under his leadership at the university.

There was a significant shift in energy and direction in 2015 within the IIC and IPC which had massive implications for the policy process with the SSDF, the work of the IIC in developing a SIP and supporting research within each of the committee's working groups. Changes in leadership at both SU and SM drew energy away from the IIC; political pressure also had negative effects on the image of the IIC in the Council. In April 2015, IIC meetings were suspended in order to allow for necessary internal discussions to take place so as to establish the most favourable positioning of the REMF collaboration, taking into consideration the leadership agenda of the newly appointed Rector, Professor Wim de Villiers.

The previous section demonstrated Stellenbosch's ULL as a learning agora operating in a fluid and contested context that while enabling experimental and innovation around complex social challenges also left the initiative susceptible to political and institutional pressure to which it was so closely tied. Since the IIC was suspended in mid-2015, interaction in the REMF has been limited. Local government elections have marked a shift in leadership at the municipality as has the appointment of a new Rector at the university. Nonetheless, in line with the strategic commitment to "community interaction" at the university, the REMF remains a recognised, albeit non-functioning entity.

4. Conclusion

This chapter set out to present the foundations of collaborative governance in Stellenbosch and to highlight the conditions that allowed for the emergence and evolution of a dynamic ULL during the period November 2013 to April 2015. A commitment to practicing transdisciplinary research and pioneering a unique approach to research partnerships that support the conduct of scientific research *with*, rather than *for*, society opened up novel opportunities for innovation and experimentation to emerge from the REMF. As an evolving governance arrangement, the REMF was not a formal or institutionalised initiative but rather operated as a legitimate holding space for cultivating more meaningful cooperation between the institutions. As such, this hybrid space became a "learning agora" for exploring emergent transdisciplinary design processes. Although it is common within sustainability science circles to emphasise the need for stakeholder participation to generate real-world solutions, there is an inadequate understanding that this will mean breaking away from traditional "extractive" research modes that define researchers as the generators of solutions and societal actors as the consumers of these solutions. As discussed in this chapter, this will require researchers to develop new skills as process facilitators and they will need to appreciate abductive modes of inquiry that allow problem statements to be jointly formulated by researchers and societal actors. The result would be shift away from systems knowledge about "what exists" and target knowledge about "what should exist", to transformation knowledge about "how things can change". This case study demonstrates that this shift is possible if emergent transdisciplinary design processes are appropriately facilitated.

References

Burns, M., Audouin, M. and Weaver, A. (2006). 'Advancing sustainability science in South Africa: Commentary'. *South African Journal of Science, 102*(9 and 10), 379–384.

Cooperrider, D.L. and Srivastva, S. (1987). 'Appreciative inquiry in organizational life'. In W.A. Pasmore and W. Woodman (Eds.), *Research in organizational change and development* (pp. 129–169).

Davies, M. (2016). 'Intermediaries and learning in sustainability-oriented urban transitions: A transdisciplinary case study from Stellenbosch Municipality'. Unpublished Master's thesis, Stellenbosch University, South Africa.

Evans, J., Karvonen, A. and Raven, R. (2016). *The experimental city.* London: Routledge.

Ewert, J. 2013. Opinion editorial: 'Poverty and inequality in Stellenbosch – the key role of education'. *Cape Times* (Cape Town). 5 February. Retrieved from http://blogs.sun.ac. za/news/2013/02/06/opinion-editorial-poverty-and-inequality-in-stellenbosch-the-key-role-of-education/

Hadorn, G.H., Bradley, D., Pohl, C., Rist, S. and Wiesmann, U. (2006). 'Implications of transdisciplinarity for sustainability research'. *Ecological Economics*, *60*(1), 119–128.

IIC. (2014). *Quo Vadis document.* Stellenbosch, South Africa: Stellenbosch Municipality.

Jahn, T. (2008). 'Transdisciplinarity in the practice of research'. *Transdisziplinäre Forschung: Integrative Forschungsprozesse verstehen und bewerten.* Campus Verlag: Frankfurt/Main, 21–37.

Kajikawa, Y., Tacoa, F. and Yamaguchi, K. (2014). 'Sustainability science: the changing landscape of sustainability research'. *Sustainability Science*, *9*(4), 431–438.

Kauffman, J. and Arico, S. (2014). 'New directions in sustainability science: Promoting integration and cooperation'. *Sustainability Science*, *9*(4), 413–418.

Leitner, H. 2012. 'Spaces of encounters: Immigration, race, class, and the politics of belonging in small-town America'. *Annals of the Association of American Geographers*, *102*(4), 828–846.

Longhurst, N. 2015. 'Towards an "alternative" geography of innovation: Alternative milieu, socio-cognitive protection and sustainability experimentation'. *Environmental Innovation and Societal Transitions*, *17*, 183–198.

Nicks, S. 2012. 'Spatial planning – Planning a sustainable Stellenbosch'. In M. Swilling, B. Sebitosi and R. Loots (Eds.), *Sustainable Stellenbosch: Opening dialogues* (pp. 24–30). Stellenbosch, South Africa: AFRICAN SUN MEDIA.

Pereira, L., Karpouzoglou, T., Doshi, S. and Frantzeskaki, N. (2015). 'Organising a safe space for navigating social-ecological transformations to sustainability'. *International Journal of Environmental Research and Public Health*, *12*(6), 6027–6044.

Pohl, C., Rist, S., Zimmermann, A., Fry, P., Gurung, G.S., Schneider, F., . . . U. Wiesmann 2010. 'Researchers' roles in knowledge co-production: experience from sustainability research in Kenya, Switzerland, Bolivia and Nepal'. *Science and Public Policy*, *37*(4), 267–281.

Polk, M. 2014. 'Achieving the promise of transdisciplinarity: A critical exploration of the relationship between transdisciplinary research and societal problem solving'. *Sustainability Science*, *9*(4), 439–451.

Regeer, B.J. and Bunders, J.F.G. (2009). *Knowledge co-creation: Interaction between science and society. A Transdisciplinary Approach to Complex Societal Issues.* The Hague: Advisory Council for Research on Spatial Planning, Nature and the Environment/Consultative Committee of Sector Councils in the Netherlands [RMNO/COS].

Reyers, B., Roux, D.J., Cowling, R.M., Ginsburg, A.E., Nel, J.L. and Farrell, P.O. (2010). 'Conservation planning as a transdisciplinary process'. *Conservation Biology*, *24*(4), 957–965.

Schauppenlehner-Kloyber, E. and Penker, M. (2015). 'Managing group processes in transdisciplinary future studies: How to facilitate social learning and capacity building for self-organised action towards sustainable urban development?' *Futures*, *65*, 57–71.

Schneider, F. and Rist, S. (2014). 'Envisioning sustainable water futures in a transdisciplinary learning process: Combining normative, explorative, and participatory scenario approaches'. *Sustainability Science*, *9*(4), 463–481.

Scholz, R.W., Mieg, H.A. and Oswald, J.E. (2000). 'Transdisciplinarity in groundwater management – Towards mutual learning of science and society'. In Belkin, S. (Ed.), *Environmental challenges* (pp. 477–487). Dordrecht, The Netherlands: Springer.

Stellenbosch Municipality. (2016). *Integrated development plan 2016/2017*. Retrieved from www.stellenbosch.gov.za/documents/idp-budget/2016-6/drafts/3429-idp-2016-17-draft/file

Stellenbosch University. (2010). *A Strategic Framework for the turn of the century and beyond*. Retrieved from www.sun.ac.za/english/Documents/Strategic_docs/statengels.pdf

Stellenbosch University. (2012). *Stellenbosch University Institutional PLAN: 2012–2016*. Retrieved from www.sun.ac.za/english/Documents/Strategic_docs/InstitusionelePlan_e.pdf

Swilling, M. (2014). Personal interview. 17 February, Stellebosch, South Africa.

Swilling, M. and Annecke, E. (2012). *Just transitions: Explorations of sustainability in an unfair world*. Tokyo: United Nations University Press.

Swilling, M., Sebitosi, B. and Loots, R. (Eds.). (2012). *Sustainable stellenbosch: Opening dialogues*. Stellenbosch, South Africa: AFRICAN SUN MEDIA.

Valentine, G. (2007). 'Theorizing and researching intersectionality: A challenge for feminist geography'. *Professional Geographer*, *59*(1), 10–21.

van Breda, J. and Swilling, M. (Forthcoming). 'The guiding logics and principles for designing emergent transdisciplinary research processes: learning experiences and reflections from a South African transdisciplinary case study: 2011–2016'. *Sustainability Science*.

Voytenko, Y., McCormick, K., Evans, J. and Schliwa, G. (2015). 'Urban living labs for sustainability and low carbon cities in Europe: Towards a research agenda'. *Journal of Cleaner Production*, *123*, 45–54.

Wittmayer, J.M. and Schäpke, N. (2014). 'Action, research and participation: roles of researchers in sustainability transitions'. *Sustainability Science*, *9*(4), 483–496.

7

BRINGING URBAN LIVING LABS TO COMMUNITIES

Enabling processes of transformation

Janice Astbury and Harriet Bulkeley

1. Introduction

As researchers have begun to trace the emergence, dynamics and nature of urban experimentation for sustainability, the role of community organisations and grassroots innovations has come to the fore (Borasi and Zardini, 2008; Seyfang and Haxeltine, 2012; Seyfang and Smith, 2007). Grassroots innovation generally occurs at a local scale and with limited financial resources. Often motivated by a belief in the value of community involvement in creating solutions to specific local problems (Seyfang and Smith, 2007, p. 591), grassroots innovations are characterised by lack of emphasis on profit, willingness to take risks (as implicit in the quest for radical alternatives), and strong underlying values (Martiskainen and Heiskanen, 2016; Seyfang and Longhurst, 2016). Grassroots initiatives thus have the potential to favour experimentation and provide a protected space in which transformation can begin to occur (Seyfang and Haxeltine, 2012, p. 382) and yet they remain inadequately embraced as a source of innovation (Hossain, 2016, p. 974). This chapter seeks to explore experimentation and innovation at the grassroots and the extent to which this allows certain community initiatives to be viewed as urban living labs (ULL) while also investigating how seeing a community initiative as a ULL helps to understand processes of transformation.

At first glance, the grassroots seem an unlikely place to find a "laboratory" such is the association of the term in everyday talk with elite forms of science and technology development. Yet, with its roots in the etymology of laboratorium, translated as "a place for labour or work", it is productive to open up the notion of the "laboratory" to understand it as a place that is set aside for the conduct of particular kinds of work. As set out in the introduction to this book, the kinds of work conducted within a ULL can be of different kinds, from serving as a test bed, a specific enclave for interventions, a showcase for the potential for alternative forms of urbanism, and a platform through which different elements of the city can be integrated in novel ways. Not all such work is explicitly cast in the terms of experimentation or laboratory interventions. Yet given similarities in the ways in which such work is carried out and its potential implications, our research sought to adopt a wide definition of ULL that included not only those ULL that defined themselves explicitly in this way but also those that bore the hallmarks of this form of urban governance intervention. Case studies of community-based ULL were therefore among those developed within the Governance of Urban Sustainability Transitions (GUST) project, which was funded by JPI Urban Europe.

One such case included in the GUST project was that of Manor House PACT (Prepare, Adapt, Connect, Thrive). It provides a good example of the particular capacity of grassroots ULL to focus on engaging the local community in a process of collective learning about the issues they face and how best to respond, and one where scaling involves broader and deeper engagement of stakeholders. Located in inner-city London, Manor House PACT was a project that sought to develop community-based and community-led responses to urban sustainability challenges. In this chapter, we first introduce the work of PACT and reflect on the ways in which it served to constitute a form of ULL in which a place-based community became the arena within which multiple approaches to sustainability designed to engage excluded communities were developed and tested. Seen from this perspective, we can think of PACT as a form of ULL that serves both as an enclave of sustainability and as a test-bed for new solutions, but that does so not through interventions that work "top down" but instead through an array of techniques that work iteratively, adjusting to experiences and adapting the forms of intervention for working with different constituents in real-time. In the second part of the chapter, we turn to focus on the processes through which PACT has sought to extend its influence beyond its boundaries, particularly focusing on the ways in which the constitution of PACT as part of a wider programme of community initiatives has enabled it to circulate, reaching into new arenas, and how its approach to learning has been central in fostering its transformative potential.

2. Making resilience in Manor House

Manor House is an area of North London, which straddles the London boroughs of Hackney and Haringey, an area known for its vibrancy and diversity, but which also faces challenges of inequality and economic, social and environmental sustainability. Like many parts of London, the area is a target for regeneration activities that seek to improve the quality of housing and public realm while encouraging economic development through public–private investment that aims to attract new residents and businesses to the area, and result in new employment opportunities. Increasing population density is seen as a contribution to meeting housing needs, economic development and enhancing sustainability, which is also targeted through the introduction of green technologies into the new buildings and more eco-friendly management of green spaces. The dominant approach to regeneration focuses largely on economic and technological inputs but community organisations in Manor House have been able to respond to the generally top-down initiatives with an empowering approach that supports local residents to play an active role in the transition process. The Manor House PACT programme has been a key component of this endeavour. It ran from February 2013 until January 2016 with funding from the UK Big Lottery Fund, which is responsible for distributing 40 per cent of all funds raised by the National Lottery. In the next section, we describe the history of the programme, the ways in which it sought to develop responses to the challenges of addressing urban sustainability in an area marked by social and economic exclusion, and which is currently undergoing regeneration, and how analysing the programme as a form of ULL provides insights into this phenomenon.

From this overview, we find that the PACT partners can be understood to have conducted two sets of experiments. First, they tested approaches to engage a community perceived as "difficult to reach" and "not interested in environmental issues" around the issue of climate change and then to empower them to take action for sustainability. Second, in undertaking the oft-neglected social side of urban regeneration in a way that is specifically focused on sustainability with respect to both (a) ensuring the benefits of physical changes and new technologies are maximised for people and environment, and (b) creating local jobs within an emerging green economy.

2.1. The origins of PACT

The Big Lottery's Communities Living Sustainably (CLS) programme was designed to support "a range of different communities, including vulnerable groups, living in geographically diverse settings to tackle the impacts of climate change focusing on social, economic and environmental challenges."[1] Central to this initiative was the incorporation of the idea that communities could and should be involved in the design and implementation of local sustainability transitions, drawing on the ideas and principles that had been developed in the United Kingdom through the Transition Town Network (Hopkins, 2011; Transition Network, n.d.). The CLS programme was explicitly experimental in seeking to foster sustainability, taking what they termed a "test and stretch" approach through which they sought to develop a programme of pilot communities where the goals that they hoped to foster could be tried out in specific local contexts. The lessons and outcomes could then be shared with relevant actors within communities, NGOs and government and used to orient Big Lottery's future work – and potentially influence the focus of other funders/government policy. In short, CLS sought to establish a collection of community-led living laboratories in which the notion of sustainability transitions could be explored. This was a process of: "trying to understand patterns, themes and lessons for the implementation of community scale initiatives. These lessons are expected to be used to help structure and develop future community scale funding programmes" (Manor House Development Trust et al., 2015, p. 1).

The local Transition Town group – Transition Finsbury Park – found the programme and its initial call for ideas to be closely aligned with their approach and were encouraged by Hackney Council to consider an application. As a volunteer run organisation, Transition Finsbury Park felt they lacked the institutional capacity to convene and manage a partnership of this kind and approached the Manor House Development Trust, an organisation formed in 2007 as part of the regeneration of the Woodberry Down housing estate. Community development trusts incorporate a community-led asset-based approach following a model espoused by the community development trust movement (that now comes together within a national network called Locality). Locality argues that its members seek to serve as anchor organisations and are marked out by their

> sense of ambition for their local neighbourhood, an enterprising approach to finding local solutions to local problems, and a clear sense that this activity should be community-led and based on self-determination. Key to sustaining [their] members in the long-term are community ownership of assets, community enterprise and service delivery.
>
> (Locality, 2015, p. 2)

This is an accurate description of Manor House Development Trust, which is an independent, resident-led Trust originally formed as part of the exit strategy for the Woodberry Down and Stamford Hill Single Regeneration Budget Partnership which ended in March 2007. Constituted as a charitable social enterprise, the Trust's role is to ensure the local community are able to benefit from the £1 billion investment into the area (Manor House Development Trust, n.d.).

MHDT (Manor House Development Trust) took the lead in submitting a bid to the CLS programme for a development grant to convene partners and prepare a full bid for funding to support a new approach to sustainability in the area, bringing together partners from Hackney Council, Haringey Council, the local traders association in Green Lanes and Transition Finsbury Park. The Manor House PACT project area was defined to include four local electoral wards that are among the most deprived wards in London (Brownswood and New River in the London Borough of Hackney and Seven Sisters and Finsbury Park in the London Borough of Haringey) (Manor House Development Trust, 2012, p. 24). Inclusion of the Woodberry Down estate was seen as particularly important because it had been a troubled social housing estate (following the path of many similar estates built in the 1960s) which in 2009 began undergoing a 20-year programme of regeneration.[2] The regeneration project is led by Hackney Council in partnership with Berkeley Homes (a property developer) and Genesis Housing Association, both of which were Manor House PACT partners. It is one of London's largest current regeneration projects and involves demolishing 1,980 homes on the estate and building more than 5500 new ones, with 41 per cent for social renting and shared ownership,[3] as well as the creation of a range of new facilities, including three new public parks.

MHDT and their partners were successful with their initial bid to develop a full proposal for the CLS programme which would focus on how transition approaches to sustainability could be developed and extended to work within the context of an area facing significant social and economic challenges as well as an extensive programme of physical regeneration (Manor House Development Trust and partners, 2012). Along with the other 29 successful bids, MDHT and their partners received a grant of £10,000 and the support of a CLS consultant "enabler" to develop their project. This was critical because it "allowed for partners to proceed with confidence knowing that there were some resources there and to develop a really rich and detailed project proposal" (Interviewee 1, November 2015). With this initial funding, MHDT hired a bid writer and funded a community survey to identify community concerns. They convened meetings to bring together key local actors who were seen as potential partners for the project. Among other things, they discussed how the outcomes identified as the goals for the overarching CLS programme[4] translated to their context. From the evidence gathered within their community and stakeholders, they developed a number of outcomes specifically for Manor House, which were grouped within three thematic strands: Home, Open Spaces and Green Vocational Training, which corresponded to needs identified in the area, including fuel poverty, worklessness and lack of positive engagement with decision makers about open spaces in the area (Interviewee 1, November 2015). The CLS enabler helped them to gather information within eight sustainability categories for the Community Sustainability Assessment Tool (CAT) that had been developed for CLS by the Building Research Establishment (BRE).

The design of Manor House PACT as a community-based "living lab" was therefore based on this extensive work of convening, pulling different actors together, engaging the community in establishing the agenda and configuring the area as one in which urban sustainability represented a central element for regeneration. This was not the work of one organisation alone, but rather a capacity that was created by the drawing together of multiple different actors:

> At that time we had two of the key partners, one was London Sustainability Exchange (LSx), which subsequently left the project, and Groundwork London, and they were really quite important in development of the proposal because their expertise in sustainability in environmental projects was really useful to bring together this coalition of people who were pulling in different ways and build support for a coherent project that was linking the different elements and finding a way to turn this Doctor Frankenstein into something that actually works. The engagement of borough level partners was also important at that stage and I know there were two key officers, one in Haringey and one in Hackney who again attended all the project development meetings and they made visits to the key project partners like Transition Towns [Transition Finsbury Park] to see the value of what was

happening already and recognising how this project could upscale what was happening already.

(Interviewee 1, November 2015)

Reflecting on the ways in which grassroots sustainability niches develop, Seyfang and Haxeltine (2012) remind us of the argument by Kemp, Schot and Hoogma (1998) that there are three key processes for

> successful niche growth and emergence: managing expectations, building social networks, and learning. Expectation management concerns how niches present themselves to external audiences and whether they live up to the promises they make about performance and effectiveness. To best support niche emergence, expectations should be widely shared, specific, realistic and achievable. Networking activities are claimed to best support niches when they embrace many different stakeholders, who can call on resources from their organisations to support the niche's growth.
>
> (Seyfang and Haxeltine, 2012, p. 384)

In the context of Manor House PACT we see that the ways in which partners were assembled who could help to create a coherent vision that could be communicated more widely, as well as partners who could connect them to broader networks, was helpful with respect to both managing expectations and building social networks. In this sense, the design of PACT as a ULL supports the importance of these kinds of intermediation in creating a space within which experimentation can take place.

2.2. Putting PACT into practice

In February 2013, the PACT partners were successful in securing funding to become one of the 12 CLS communities. Responsibility for the development and implementation of the programme was distributed among six delivery partners: Transition Finsbury Park, Groundwork London, London Wildlife Trust, Genesis Housing Association, and London Sustainability Exchange, which was later replaced by a small consultancy called Green and Castle. There were also five strategic partners including the two London boroughs associated with Manor House and their corresponding housing associations, Hackney Homes and Homes for Haringey, along with the property developer Berkeley Homes (a key actor in regeneration of the Woodberry Down area of Manor House). At the heart of PACT was a four-fold focus: prepare, adapt, connect, thrive:

> The partnership came up with the acronym PACT as it clearly communicates the different stages and principles it wants to use in approaching the issue of climate change. The partnership's vision is for people living in our area to

thrive whatever might happen as a result of climate change; to achieve this will involve local people preparing, adapting and connecting with each other.

(Manor House Development Trust and partners, 2012, p. 41)

In conceiving of urban sustainability in this way, the programme sought to strengthen social networks and empower the community to live more sustainably by managing natural resources such as energy, water and food wisely; by creating new economic opportunities for local people within an emerging green economy and through improving physical, emotional and mental health. The programme provided a means through which to experiment and adapt different kinds of interventions that could meet these broad intentions. This vision for the ways in which urban sustainability might be translated into the concerns of socially and economically disadvantaged groups was in turn translated initially into the three thematic strands, Home, Open Spaces and Green Vocational Training, which reflected needs identified in the area. As the programme developed, strong local interest in activities related to "Growing and Eating" served to create the impetus for the development of a new thematic strand (see Table 7.1).

This interest in gardening and outdoor spaces reflects a global trend where urbanising populations seem to feel an increasing need to get their hands dirty and reconnect with nature and sources of sustenance (Fuller and Irvine, 2010). Such contact is known to have significant health and well-being effects (Guitart, Pickering and Byrne, 2012; Turner, Henryks and Pearson, 2011). Opportunities for such contact are particularly important in areas like Manor House where many residents rely on local public outdoor spaces for such benefits (Guitart, Byrne and Pickering, 2012) and where the need will become greater as the regeneration process densifies the neighbourhood with the local population more than doubling over the coming years. In this context, PACT will leave an important legacy through community gardens, increased interactive activities in Finsbury Park and other green spaces, and transformation of the reservoir, now known as Woodberry Wetlands, a new nature and wildlife reserve developed and managed by PACT partner London Wildlife Trust.

Not all of the activities undertaken by PACT were new and instead built on previous work and existing infrastructure. For example, MHDT had initiated a community garden and community meals, and trained and supported Health Champions under the Lottery funded Well London programme. Well London, a partnership initiative with the Mayor of London, ran from 2008 to 2015 and sought to help local communities improve their health and wellbeing. MHDT managed Well London within North East Hackney focusing particularly on supporting Health Champions who disseminated their own learning about health promotion, first aid, money management, food hygiene, and cookery and gardening. The programme resulted in capacity development as well as the introduction of new activities (e.g. fitness, yoga, cycling lessons, community choir, cooking and eating together) and transformation of local spaces (Manor House

TABLE 7.1 Manor House PACT activities organised by thematic strand

Home (and school)	Open spaces	Growing and eating	Green vocational training
– Home energy visits (Groundwork London) – Closer neighbours (energy efficiency and sustainable practices) (MHDT) – Building conversations (building social capital and addressing fuel poverty and other vulnerabilities) (Green and Castle) – Climate change and sustainability education in schools (new strand) (Transition Finsbury Park)	– Hidden river festival (most partners, including Hackney Council and Berkeley Homes with the latter now likely to take charge of future events) – Wildlife Walks (London Wildlife Trust) – Bushcraft training (for youth centre and broader community) (London Wildlife Trust) – Foraging walks (cross-cutting theme with "growing and eating") (Transition Finsbury Park)	– PACT meals (Transition Finsbury Park with various partners hosting the meals) – Gardening clubs at various locations including the Redmond Community Centre and local housing estates (Transition Finsbury Park in collaboration with partners at sites) Permaculture	– Volunteering (MHDT) – Green Jobs Training – Energy assessors and green construction (MHDT) – Woodberry Wild Talent (ecological land management) (London Wildlife Trust + Genesis Housing) – Support for new green businesses (Genesis Housing) – Creating a forest garden + training (Transition Finsbury Park)

Development Trust, 2016a, pp. 15–19). Within PACT, MHDT and partners expanded the gardening and meals activities (bringing in new partners and working at a broader range of sites) and linked them to concerns and learning opportunities related to energy, waste and climate change. They brought in new food related activities such as foraging and developed training and employment opportunities (some of which were offered by Transition Finsbury Park expanding its previous activities in local green spaces), which augmented the employment development that MHDT was already undertaking in relation to construction work within the regeneration project. This continued under PACT with a greater emphasis on green construction. To these were added other new opportunities to develop skills in green space management led by London Wildlife Trust in collaboration with Genesis Housing Association.

Volunteer Energy Champions were trained to provide their neighbours with advice and support to keep warm and save energy. Lessons learned from training Health Champions were applied, such as building on existing networks of relationships, knocking on doors and going block by block building trust as they went, and empowering volunteers to take on leadership roles and find ways to continue and further develop services. Energy Champions sometimes accompanied "Green Doctors" on PACT home energy visits in order to provide support and follow up with residents. The Green Doctor is a successful model developed by PACT partner Groundwork which involves sending a domestic energy expert to assess and offer solutions to home energy needs.[5] PACT provided the opportunity to test a two visit model (one visit is the norm within Groundwork), which was found to be effective with respect to ensuring that people understood how to use their new energy efficiency technologies and to support them in maintaining energy and water saving practices to which they had committed during the first visit.

In pursuing these activities, PACT used two important sets of techniques – those through which it came to *know* the urban arena within which it was operating and those through which it came to *engage and enrol* a host of actors and entities into the work of the programme. Coming to know the urban arena required not only that those engaged in the intervention gained an understanding of the place through which they were seeking to further urban sustainability, but also that they ensured that traditionally "hard to reach" communities came to know sustainability in a way that made it meaningful to them. PACT activities had a strong focus on bringing people together to share information and motivate one another. This allowed it to facilitate social learning[6] about things like energy saving, gardening, healthy eating, local nature, etc. PACT also offered a range of training activities, for example forest gardening, retrofitting, energy champion, energy assessor, which promote sustainable action and facilitate access to employment. This was not a matter of "knowing" that often accompanies the development of ULL, with a focus on the monitoring or sensing of the urban realm, but instead of bringing sustainability to "where people are" (which often means knocking on their doors) and focus on building relationships.

One activity that contributed to both knowing and engaging, while also serving as an evaluation tool, was the Impact Measurement Framework developed for PACT by CAN Invest (a consultancy specialised in social impact). It included pre- and post-activity data collection using surveys and forms. As well as assessing effects, the forms served to build a database of community members and their needs and interests in order to be able to tailor activities and make sure that relationships were maintained with everyone who came into contact with PACT. Relationship building was a key aspect of engagement and the extent to which PACT was successful in maintaining and deepening relationships was assessed through interviews and focus groups undertaken at the end of the PACT programme. These tracked the "journey" of PACT participants through a series of activities and ideally a process of engagement and empowerment.

The techniques of knowing, engaging and enrolling were also practiced within the living laboratory through bringing sustainability into people's lives in a meaningful way. This revolved around both seeking to transform everyday practices in ways that enhanced sustainability, and pursuing a series of activities intended to support the creation of the green economy as a means through which to foster employment, development and regeneration. Transformation of practices is understood within PACT to be a long and step-by-step process, likely to begin with simple things like the very successful PACT meals. Engagement efforts need to address people at whichever stage they are in the process. As one interviewee pointed out, behaviour change is like taking up foraging:

> Foraging is quite a good model for behaviour change because you have the entire progression. People are at different stages along the whole timeline of having an idea on your periphery vision as it were, the concept that there might be edible plants around you, through to being able to recognise that plant, to then actually taking the plunge and eating something. People are often quite resistant to that, especially if it's a flower [laughs]. And then maybe being more adventurous, maybe picking something on a foraging walk and taking it home and cooking it and trying it out, making something like a soup. To then maybe modifying their behaviour so that they're actually going out of their way [saying to themselves]: "Oh, it's that time of year when this plant is in abundance, I'm going to go and harvest it. It takes a long time to get there and the main reason why you wouldn't go along that continuum is lack of need." So until people actually need to eat foraged foods, it's just something you will do if you really want to, if you like the taste of the food and if you just want to not spend money on buying it. But people can turn up who are at any point along that line and they will ask different kinds of questions. It's the same as other behaviour changes.
>
> (Interviewee 4, November 2015)

In seeking to build a green economy, individuals were engaged in both training and volunteering programmes that sought not only to re-orientate their practice, but to shift the networks of which they were part. This proved to be a successful approach. Some training programmes piloted through PACT have been turned into accredited courses, including the first accredited forest gardening course in the United Kingdom. Four cohorts (54 students) have completed the course and a social enterprise has been created to deliver it in future. A number of alumni have found jobs and in some cases have found opportunities to introduce the approaches they have learned into their new places of employment. The volunteering programme was shown to have increased engagement, health and well-being and access to employment. It deployed volunteers across PACT activities and a key factor in its success was its very personalised approach to recruiting, placing and supporting volunteers, understanding what sort of support they required and how to frame the opportunities in ways that would attract them.

This overview of the broad range of PACT activities makes evident the presence and significance of the techniques of knowing, engaging and enrolling. The PACT partners set out to deeply know their community and to foster knowing about sustainability within the everyday practices and concerns of community members. This knowledge and the activities that facilitated the associated learning, which were explicitly situated in a process of relationship building, served to engage a broad range of community members in the PACT project and to enrol some of them as full partners in sustainability transition.

2.3. PACT as a laboratory for experimenting and learning

Manor House PACT therefore differs from many ULL in terms of the ways in which it seeks to practice experimentation and learning. Rather than being based on systemic or formalised processes of evaluation or learning per se, though these were certainly part of how the project was developed, it worked by learning from and with the communities with which it was engaged. It became a safe space in which new ways of thinking about urban regeneration for these communities could be explored and put into action through a variety of means, even if these were quite familiar in other urban contexts, which in turn served as a process of "learning by doing" for the stakeholders involved and in turn informed the CLS goal of testing and stretching what it was possible to achieve in terms of community-led sustainability. Here, experimentation is a process of adaptive trial and error (Cook, Casagrande, Hope, Groffman and Collins, 2014).

In the case of PACT, a core premise was that the "hard to reach" can be reached and engaged in relation to sustainability issues. The Manor House area served as the laboratory in which this hypothesis could be tested and more importantly from the perspective of the PACT partners where multiple approaches to reaching and engaging local communities could be tried and tested in order to identify effective approaches that could continue to be applied in the course of the PACT project and in the future. These included working with schools and children in special needs education to communicate climate change, taking a face to face approach to engage diverse ethnic communities and the development of a programme of home energy visits (modelled on the Green Doctor model) through which to engage residents with changes to everyday practices, and using community lunches as a means through which to engage participants in collective action in response to issues such as how to reduce energy bills.

This drive for experimentation came not from seeking to optimise sustainability per se, but through a strongly held commitment that all parts of society could be engaged with this agenda so long as the approach to engage them was suitable. Experimentation was then needed to ensure that sustainability reached the right people, rather than in order that people would reach the right kind of sustainability. This was particularly evident in the case of the ambition to create access to jobs in an emerging green economy. A number of organisations in North London were trying to facilitate access to green jobs for local people and complaining that the anticipated jobs did not exist. This was attributed to factors such as uncertainty in government policy related to low carbon transition and to procurement practices that favoured large companies for whom recruiting local people in need of jobs was not a priority.

While these complaints were justified, PACT did not give up and tried successive "experiments", including targeting different kinds of jobs, changing training and volunteering opportunities and the ways of approaching employers. In each case project partners analysed what had worked and not worked, what was the root problem, and how to adjust their strategy. When the problem was at a policy level, they communicated this to the local authorities concerned and/or to the CLS "learning partners" which were organisations with links to national policy actors.[7] The PACT partners learned lessons about how to ensure that training led to jobs, particularly with respect to engaging employers from the outset by asking them to agree on the content, contribute to the cost and guarantee interviews for people who completed their training. MHDT staff hoped to apply this learning in its future efforts to help local people to access employment created by regeneration activities.

This ambition to experiment in order to find a way to overcome a specific challenge – of engaging excluded communities in the sustainability agenda and enabling these groups to benefit from the potential of the green economy – was coupled with a condition in which experimentation was encouraged by the funding organisation. The CLS programme explicitly sought to encourage a "test and stretch" approach, in which communities were encouraged to test ideas and approaches, to share learning among local actors and the broader CLS network, and to expand effective activities or seek alternative ways to achieving their desired outcomes. "Test and stretch" was a general approach for the whole programme and also an invitation to try out specific ideas for which funding and support were made available. For example, Manor House PACT asked for additional funding to assess the feasibility of developing solar PV generation in the neighbourhood and was assisted by the Energy Saving Trust that was one of the organisations in the CLS Learning Partnership. It took PACT some time to gain the confidence that there was genuine scope for trial and error through this programme, but once this had been established those involved in implementation recognised that they had been given a lot of freedom to try things out and then apply the learning:

> if it was couched as "this is a training offer, this is something where we can come along, do it all for you, there's no administration. All you have to do is come along and talk to us, give us your time, spend some time with your neighbours, colleagues, and that was a much easier mechanism for us to be able to reach people.
>
> (Interviewee 1, November 2015)

One of the PACT delivery partners pointed out that the framework provided by the funder facilitated experimentation and learning and that the characteristics of people who were delivering the project were also important:

> From the start, the way that the project was framed to us from Big Lottery was that it was experimental and we are trying stuff out and we knew that

it was important for them that whatever we do can then be replicated elsewhere. They're not interested in tailored, bespoke, one off projects. They're interested in stuff that can be copied by others and that other people can learn from and the people [working with PACT] have gone into it with a genuine spirit of exploration, and it couldn't have worked otherwise – it would have just been a waste of time.

(Interviewee 4, November 2015)

This indicates both the importance of a supportive context, in this case the approach of the funder, and the capacity and willingness of key actors, such as the PACT delivery partners, to seize the opportunity to experiment. A clearly articulated goal, in this case, "reaching the hard to reach", and determination to reach it served to frame and sustain the process of experimentation. PACT leaders played important roles in articulating the vision and communicating confidence in the capacity of the community to achieve the desired outcomes in the manner described by Martiskainen (2016) as supportive of development of grassroots innovation. The commitment to the values of inclusion and engagement, which characterises this and other grassroots ULL, created a patient and supportive environment that provided a protected space for a transformative process to unfold.

3. Enabling transformation: Fostering processes of change

Beyond understanding the design and practice of ULL, an understanding of their potential and consequences requires that we engage with the processes through which they (more or less intentionally) have transformative effect. Transformation in this case refers to lasting changes in people's ideas and practices, as well as tangible changes to the place they inhabit and to its social structures and activities. What is evident from the example of Manor House PACT is that ULL often emerge from processes of transition that have been started in other contexts. In short, rather than starting from a "blank slate" and fostering discrete processes of transition towards sustainability, we find that Manor House PACT is more appropriately thought of as part of a wider "ecology" of urban experimentation (McGuirk, Bulkeley and Dowling, 2014), in which it both serves as an arena in which the transformative potential of sustainability transitions learnt elsewhere is *translated*, and as a domain within which new dynamics of *learning, scaling and empowering* are fostered.

In the case of PACT, the process of *translating* – where visions and practices for sustainability transition in one context are interpreted in another – began from the outset of the design and constitution of the ULL. The conception of the CLS programme was informed by a scoping report concerning how communities might respond to climate change, which drew heavily on the notion of Transition Towns. The funding programme itself then served to translate these principles and practices into a range of sites, not only through encouraging bids that took these issues on board but specifically in the case of short-listed projects providing direct support to enable the translation of these principles into their own contexts. In particular, the neighbourhoods in which Manor House PACT sought to intervene represent quite a different context from those within which Transition Towns have traditionally been developed and they had to significantly adapt the approach. For example, they used paid facilitators and promoted the initiative to community members in terms of opportunities to develop skills and with no obligation to take on administrative tasks. Rather than replicating a monolithic version of sustainability, the active work of translation enabled the PACT partners to gain insight from the transition approach and put it to good use in both securing resources and developing new ways of addressing the challenge of engaging hard to reach communities in sustainability.

The next and critical set of processes that have been central to the working of PACT as a living laboratory are those that can be broadly termed *learning*. As set out above, learning was at the heart of many of the practices within PACT. The programme partners were able to share their learning widely because they experienced a lot of interest in PACT. People heard about PACT through the networks to which its various partners were connected. Large-scale PACT events like the Hidden River Festival raised its profile and its variety of activities created entry points for different people. The PACT manager described how it had become a sort of "beacon" and a "lightening rod" for those interested in the broad interpretation of sustainability manifest in the project. Yet despite the emphasis on learning within the CSL programme, no specific budget was put aside for this and the work of making space and time for the kinds of informal interactions and learning that were undertaken by the project had to go alongside the day to day business of the project. Beyond the activities it organised itself to foster learning, PACT was an active participant in the learning and sharing activities of the wider CLS programme including the CLS Hub. This was an online support and learning platform managed for CLS by the Groundwork UK Learning Partnership where the learning partners (Groundwork UK, The Energy Saving Trust, The Federation of City Farms and Gardens, The New Economics Foundation and BRE) mentored CLS communities, transmitted lessons learned on the ground to policy makers and produced toolkits based on learning from the communities.[8]

The capacity of the PACT partners to "learn by doing" was enhanced by the continuity provided by the well-respected PACT Project Manager and Transition Finsbury Park Coordinator who were able to carry earlier lessons forward, provide thoughtful analysis and create a safe reflective space where all partners felt at ease in talking about what had and had not worked. The nature of their organisations was also significant with respect to channelling learning in and back out. Transition Finsbury Park embodied the Transition Towns principles in which the initiative was rooted while MHDT was a key partner in the regeneration process with relatively stable funding and occupying a welcoming space where many PACT activities took place. The framework created by these two key organisations seemed to provide a buffer for changes taking place within other PACT partner organisations. The two partners where personnel changes did have significant impact were the two Councils and particularly Haringey. Both Councils demonstrated strong commitment to the PACT project during its development and this represented a potentially important vehicle for ensuring its broader impact. There was a sense that the staff who were involved in the early stages were carrying lessons from PACT back to their Councils. This was not the case later on. The local authorities were thus less successful in sustaining the relationship and some sort of leadership role within PACT through their personnel changes than were the third sector organisations.

Alongside the processes of learning that were fostered through PACT were also forms of *scaling*. Interestingly, in the case of PACT scaling did not so much involve moving up or beyond the scale at which it had been established, but extending the activity in both space and time to incorporate more household and individual participants on the one hand, and to continue to roll out different programmes year on year even when the initial funding had ceased. Successful activities like food growing were expanded and replicated in different parts of the community. Home Energy Visits, building on their success within PACT, are continuing and in 2016 MHDT succeeded in raising funds for 200 visits and getting Hackney Council to match this with funding visits to another 200 homes. This is a form of scaling that involves *extension* and the continual enrolling of new actors, sites and finance within the community in order to maintain and embed community-based experiments. This is indeed rare within this form of experimentation, given the precarity of funding for grassroots initiatives and the "burn out" with which they are often associated.

Finally, central to the processes of transformation that PACT engendered have been forms of *empowering* (Smith and Raven, 2012). Engaging local residents, including "hard to reach" community members, was at the heart of the PACT project. The PACT approach to engagement is very personalised. It begins with meeting people where they are (starting in their own homes and through community meals hosted in different venues). Activities are designed to align with people's interests (and constantly adjusted) and when people participate in one activity, they are encouraged to participate in others. This served to create empowerment "journeys" where individuals who may have been involved in one kind of activity became more centrally engaged in other forms of sustainability practice. According to PACT's enabler, one of the most striking things is that:

> The people in Manor House have actually bought into PACT despite the fact that the majority of them are in relatively low income situations. They have really got on board in terms of making changes to their habits and lifestyles to try to mitigate or adapt to climate change. They have developed a sense of community spirit in what is a very diverse community. There is a lot of building work and transitions for families having to move out of their properties. What has been achieved in the face of all those really complicated factors is impressive.
>
> (Interviewee 3, November 2015)

Beyond seeking to empower individuals to change their everyday practices and open up the spaces of the green economy to their participation, the PACT approach assumed that people need to proactively build community during transitions. It operates in accordance with the Transition Network principle that community is what supports change. Surveys conducted for PACT showed that more frequent participation in activities was linked with proportionally greater changes in seven types of green and healthy behaviours, which included walking more, growing more food, buying seasonal and local food, increased recycling and reducing food waste, and energy and water usage (Manor House Development Trust, 2016a, p. 49). Participation in activities increases people's sense of belonging, which is important because "the sense of feeling part of a community that is working together for a shared goal has been the biggest driver of behaviour change for PACT" (Manor House Development Trust and partners, 2015a, p. 24).

PACT also recognised that while the natural environment improves (as was foreseen in Manor House), people need to know how to use and care for it in order to safeguard positive changes and maximise benefits for themselves. Without these supports, the material transition of the place is at risk of becoming an excluding and disempowering process if the community experience it as a significant change in their physical environment happening beyond their control. Such disempowering experiences can undermine the capacity of local community members to contribute to sustainability transitions – and may result in them directing their energies to resisting change of any kind.

Much of the criticism levelled at ULL is that they foster a particularly instrumental, technocratic approach to sustainability transitions (Evans and Karvonen, 2014). Perhaps nothing could be further from the kind of transformations that have emerged in PACT. Incremental, difficult to trace and often small in scale, they have focused on understanding and learning about issues of exclusion, powerlessness and inequality in order to engage different kinds of communities in sustainability transitions. In doing so, the processes that come to the fore are those of social learning and empowerment. A long-term, progressive funding scheme, focused on the development of community-led, experimental approaches to sustainability transitions created a protected space within which the commitment of the partner organisations to their local community and the determination of individuals involved in the project to reach their target audience no matter how many attempts it took could be harnessed.

4. Conclusions

Manor House PACT has functioned explicitly as a laboratory for learning, a space within which "trial and error" approaches have been welcomed, with processes of translation, learning, scaling and empowering given space to flourish from the grassroots. Yet at the same time, it has relied on the strategic intervention of a large national funding programme as well as the willingness of municipal actors to get involved with a new approach to urban regeneration. This suggests that successful grassroots forms of ULL may be those which are able to intermediate between the various capacities to engender transformation that operate at different scales – national, civic and community – and bring them together to create new kinds of practices and processes for urban change.

The fact that there was a process of transformation already under way in the form of the physical regeneration may have helped to open up a protected space for experimentation. Stakeholders understood that the regeneration process created opportunities for other sorts of transformation and also entailed high risks – including the potential for disempowerment and frustration of local residents seeing their community changed by forces beyond their control. Risk, uncertainty and feelings of disempowerment are also very present in the context of climate change. Working within this complex situation where there are no clear paths forward may have helped to open up space for experimentation while also creating the kind of enclave setting within which "laboratory" approaches emerge. It may also indicate that PACT might be a model for other urban regeneration initiatives.

The creation of a community development trust (MHDT) with responsibility for facilitating the social side of the regeneration provided the infrastructure for a multi-partner initiative like PACT to develop. It built on the regeneration partnerships that MHDT had already developed and integrated a new or stronger focus on climate change adaptation and sustainability into the existing work of a number of partners. Scaling within PACT emphasised extending its activities in both space (enrolling more participants) and time, including beyond the limited period of funding, which is an important focus for community organisations. The networks to which each of the partners belonged provided an opportunity for dissemination of learning and potential for scaling in the way that Seyfang and Longhurst (2016, p. 19) have shown to be important for grassroots innovation. Combined with PACT activities and events, these networks helped PACT become a showcase for transitioning to sustainability in a community facing social and economic challenges. CLS assisted this process by providing both forums for exchange and channels for communicating to funders, policy makers and organisations in other communities facing similar challenges.

The local scale of the PACT project and the clearly bounded area in which it unfolded made it possible for PACT partners (organisations fully embedded in the community and working in close collaboration with one another) to see the whole experiment. They were able to follow the different activities and the links among them, and observe on-going changes in place and people. CLS provided funding and other forms of support that encouraged experimentation and learning, allowing PACT partners to keep trying different approaches until they found what worked. PACT like all of the community initiatives supported by CLS was called upon to test and stretch approaches at a local level in order to test and stretch national-level strategies for supporting communities in confronting climate change and addressing sustainability issues. The importance of such a funding scheme in creating the conditions for emergence of an ULL cannot be overstated, nor can that of the capacity of the community organisations involved to embrace the opportunity for experimentation and learning.

PACT documented its learning as well as its achievements explicitly describing methods of data collection in an accessible format (see Manor House Development Trust, 2016a, 2016b). Its work has been fed into the CLS Learning Partnership process in order to analyse findings across the 12 communities, which will likely lead to further learning and scaling through a range of different channels. Here, the funder Big Lottery CLS became a learning and scaling conduit that reflected approaches developed at the grassroots through Transition Towns back to local community actors, supported them to adapt these approaches to their contexts and to learn from their experiences (where experimentation was encouraged) and share this learning across the network – and to channel it to policy makers. Big Lottery then took up the new learning from across the funded projects and communicated their intention to use this in defining their future granting strategy. In this way, Manor House PACT has shared its learning at multiple levels while always staying focused on applying this learning directly to transforming the community of which it is part, as is fitting for a grassroots ULL.

Notes

1 See: www.biglotteryfund.org.uk/global-content/programmes/england/communities-living-sustainably
2 See: www.hackney.gov.uk/woodberry-down.htm
3 Shared ownership schemes are provided through housing associations. A first-time house buyer (or former owner who can no longer afford to buy) purchases a share of the property, pays a mixture of mortgage and rent and can eventually purchase the other shares in order to assume full ownership.
4 Communities living sustainably outcomes: (1) vulnerable people affected by the impacts of climate change are able to make greener choices to help improve their quality of life; (2) communities are better prepared for environmental challenges and longer term environmental change and understand the improvements they can make to live more sustainably: (3) communities maximise the use of their assets and resources to create new economic opportunities and live more sustainably by, for example using the skills and knowledge of individuals within their community to create green social enterprises and jobs and (4) communities have a greater understanding of and more opportunities to use natural resources more efficiently.
5 Green doctors offer residents a range of simple energy efficiency measures including low energy light bulbs, draught proofing and water saving devices. They also signpost residents if eligible to government and energy company grants that can help them to install more significant energy saving measures, such as loft or solid/cavity wall insulation and boiler replacements. They offer residents debt assistance and also offer energy tariff or company switching advice. Another important part of the service is advising residents about simple behavioural changes that can help them take control of their energy use – changes like heating rooms individually, using energy saving light bulbs and washing at 30°C (www.groundwork.org.uk/Sites/london/pages/green-doctors-lon).
6 Social learning is the collective learning through action and reflection that results in enhancing a group's ability to change its underlying dynamics and assumptions (Tippett and Searle, 2005).
7 The five learning partners were Groundwork UK, The Energy Saving Trust, The Federation of City Farms and Gardens, The New Economics Foundation and Building

Research Establishment (BRE). These organisations had links to relevant policymakers and were able to communicate learning from the CLS supported projects.

8   See: www.communitieslivingsustainably.org.uk/library/.

References

Borasi, G. and Zardini, M. (Eds.). (2008). *Actions: What you can do with the city*. Montreal, Canada: Canadian Centre for Architecture.

Cook, W. M., Casagrande, D. G., Hope, D., Groffman, P. M. and Collins, S. L. (2004). 'Learning to roll with the punches: Adaptive experimentation in human-dominated systems'. *Frontiers in Ecology and the Environment, 2*(9), 467–474.

Evans, J. and Karvonen, A. (2014). ' "Give me a laboratory and I will lower your carbon footprint"! Urban laboratories and the governance of low-carbon futures'. *International Journal of Urban and Regional Research, 38*(2), 413–430.

Fuller, R. A. and Irvine, K. N. (2010). 'Interactions between people and nature in urban environments'. In K. J. Gaston (Ed.), *Urban ecology* (pp. 134–171). Cambridge: Cambridge University Press.

Guitart, D., Pickering, C. and Byrne, J. (2012). 'Past results and future directions in urban community gardens research'. *Urban Forestry and Urban Greening, 11*(4), 364–373.

Hopkins, R. (2011). *The transition companion*. Totnes, UK: Green Books.

Hossain, M. (2016). 'Grassroots innovation: A systematic review of two decades of research'. *Journal of Cleaner Production, 137*, 973–981.

Kemp, R., Schot, J. and Hoogma, R. (1998). 'Regime shifts to sustainability through processes of niche formation: The approach of strategic niche management'. *Technology Analysis and Strategic Management, 10*, 175–196.

Locality. (2015). Theory of change. Retrieved from http://locality.org.uk/wp-content/uploads/Theory-of-Change.pdf (accessed 13 November 2016).

Manor House Development Trust. (n.d.). Who we are. Retrieved from www.mhdt.org.uk/about-us/who-we-are/ (accessed 13 November 2016).

Manor House Development Trust. (2016a). *Social impact report manor house development trust 2009–2015, September 2015*. Retrieved from www.mhdt.org.uk/wp-content/uploads/2014/01/MHDT-Impact-Report-2009-2015.pdf (accessed 13 November 2016).

Manor House Development Trust. (2016b). *Manor House PACT Impact Report 2013–2016, June 2016*. Retrieved from www.mhdt.org.uk/wp-content/uploads/2016/08/PACT-FINAL-Report.pdf (accessed 13 November 2016).

Manor House Development Trust and partners. (2012). Manor House PACT Project Delivery Plan. Bid submitted to Communities Living Sustainably, Big Lottery Fund.

Manor House Development Trust and partners. (2015a). Manor House PACT year 2 interim learning report, May 2015. London: MHDT.

Manor House Development Trust and partners. (2015b). *End of project CAT review workshop – Briefing Paper, November 2015*.

Martiskainen, M. (2016). 'The role of community leadership in the development of grassroots innovations'. *Environmental Innovation and Societal Transitions*. Retrieved from http://dx.doi.org/10.1016/j.eist.2016.05.002

Martiskainen, M. and Heiskanen, E. (2016). 'Politics of grassroots innovations'. In *Second International Conference of the Sustainable Consumption Research and Action Initiative (SCORAI)*. 'Transitions Beyond a Consumer Society', 15–17 June 2016, University of Maine, Orono, ME.

McGuirk, P., Bulkeley, H. and Dowling, R. (2014). 'Practices, programs and projects of urban carbon governance: Perspectives from the Australian city'. *Geoforum, 52*, 137–147.

Seyfang, G. and Haxeltine, A. (2012). 'Growing grassroots innovations: Exploring the role of community-based initiatives in governing sustainable energy transitions'. *Environment and Planning C: Government and Policy*, *30*(3), 381–400.

Seyfang, G. and Longhurst, N. (2016). 'What influences the diffusion of grassroots innovations for sustainability? Investigating community currency niches'. *Technology Analysis and Strategic Management*, *28*(1), 1–23.

Seyfang, G. and Smith, A. (2007). 'Grassroots innovations for sustainable development: Towards a new research and policy agenda'. *Environmental Politics*, *16*(4), 584–603.

Smith, A. and Raven, R. (2012). 'What is protective space? Reconsidering niches in transitions to sustainability'. *Research Policy*, *41*(6), 1025–1036.

Tippett, J. and Searle, B. (2005). 'Social learning in public participation in river basin management – Early findings from HarmoniCOP European case studies'. *Environmental Science and Policy*, *8*, 287–299.

Transition Network. (n.d.). 'Transition towns'. Retrieved from www.transitionnetwork.org (accessed 13 November 2016).

Turner, B., Henryks, J. and Pearson, D. (2011). 'Community gardens: Sustainability, health and inclusion in the city'. *Local Environment*, *16*(6), 489–492.

8

HOMELABS

Domestic living laboratories under conditions of austerity

Anna Davies

1. Introduction

With unsustainable consumption described as the mother of all environmental issues and three domains – food, mobility and housing – responsible for more than two-thirds of greenhouse gas emissions, attention to the impacts of everyday consumption has grown dramatically over the past decade. This has led to calls, most particularly within Europe, for domestic sustainability transitions and explorations of innovative ways in which unsustainability can be disrupted and reoriented (Davies et al., 2014). Focused on different scales, sectors and issues, this work has adopted a range of conceptual framings and empirical tools to better comprehend the complexity of why people consume the way that they do "at home" and the ramifications of that consumption. As noted by Camaren and Swilling (2014), this diversity has been relatively successful at determining problems, but has been less successful in terms of thinking through solutions. Living laboratories (defined in their broadest sense, as articulated throughout this volume) have emerged in response, as sites of explicit experimentation for developing, testing and evaluating ways of living differently (Davies and Doyle, 2015a). As the landscape of these living laboratories evolves and experiences proliferate, discussions around the possibilities for trans-laboratory learning across territorial boundaries are moving to the fore. While transition theorists have provided useful conceptual frameworks to understand how transitions have occurred in the past, questions remain regarding the ability of these theories to explain current processes and, more importantly, predict potential points of intervention to influence patterns of change in the future. Equally, to date, while the field of transitions management is more empirically grounded, it has tended to rely on a limited number of case studies in particular geographical settings and socio-political cultures that have been relatively amenable to matters of sustainability in public policy. The extent to which the outputs of living laboratories in these contexts might inform responses to sustainability transitions in quite different contexts is very much a live discussion that will require a substantial body of empirical evidence drawn from a wide variety of places. Expanding the existing empirical architecture around sustainability transitions, this chapter reflects on the experience of co-designing and conducting a suite of collaborative, transdisciplinary experiments, called HomeLabs (see Davies and Doyle, 2015a; Devaney and Davies, 2016 for a full description of the method). These HomeLabs explored with householders their experiences of engaging with disruptive supports for more sustainable household consumption in Ireland; a context that was, at the time of the experiment, characterised by fiscal constraint and austerity politics rather than sustainability agendas.

In this chapter, specific attention is given to the experience of conducting and evaluating HomeLabs focused on disrupting domestic water consumption (Washing HomeLabs), which were rolled-out in parallel with substantial institutional reform of the water sector, including the formation of a new national water utility, Uisce Éireann (Irish Water) and the development of a new, and highly controversial, fiscal system for managing the costs of water treatment and supply. As such, this chapter provides a productive space to reflect on the opportunities and challenges of applying findings from research-driven, collaborative living laboratories for sustainability transitions in highly contested contexts and dynamic policy arenas.

2. Experimenting for sustainability transitions in Ireland

In Ireland, consumption in general, and household consumption in particular, rose significantly between the late 1990s and 2008, a period of economic expansion known colloquially as the "Celtic Tiger" (EPA, 2006). While levels of consumption decreased following the deep period of recession and attendant austerity measures that followed, there has yet to be an absolute decoupling of economic growth and resource consumption (EPA, 2012). Despite an active research programme in sustainability science, led by the research funding arm of the Environmental Protection Agency, sustainability issues were not at the top of policy agendas during the boom time of the Celtic Tiger economy. Since the bursting of the Celtic Tiger bubble in 2008, successive Irish Governments have held a strong line on cutting public sector pay budgets and stimulating jobs and economic growth that has narrowly defined sustainability along its economic axis. In the absence of strong support for broader sustainability transitions across Government departments, experimentation through HomeLabs was a less directed process combining diverse actors to stimulate new spaces of innovation for the testing and evaluation of alternative practices (Bulkeley and Broto, 2012). The HomeLabs began in Dublin in 2013, just as Ireland exited from the financial bailout programme administered by the Troika (International Monetary Fund, European Central Bank, European Commission) and after four years of external economic oversight and harsh internal austerity policies that saw cumulative cuts to public spending of over €30 billion. Developed with a commitment to co-design principles and derived from practice-oriented participatory (POP) backcasting activities (Davies and Doyle, 2015a), the experiments embodied a research-led, exploratory living laboratory approach. The experiments were funded under the Irish Environmental Protection Agency's STRIVE programme as part of the CONSENSUS (consumption, environment and sustainability) project, which ran from 2009 until 2016. CONSENSUS was the first large-scale, all-Ireland (including the Republic of Ireland and Northern Ireland) study of domestic consumption practices. Following critical analysis of existing tools and technologies governing consumption (Davies et al., 2010; Pape et al., 2011) – that is, the variety of ways through which consumption is managed (e.g. from taxes to information campaigns) – and a lifestyle survey (Lavelle, 2013), the substantive phase of CONSENSUS focused on four challenging areas of household consumption: mobility, heating, washing and eating (Davies et al., 2014). Within the heating, eating and washing research, a trans-disciplinary co-design approach was adopted, utilising a variant of participatory backcasting (see Davies et al., 2015). Actors from a wide variety of backgrounds and positions came together to generate regulatory, socio-cultural and technical innovations for more sustainable household consumption by the year 2050. Frameworks were then developed identifying short-, medium- and long-term interventions that might combine and align to support transitions to more sustainable practices (see Figure 8.1 for an illustrative extract of the Washing Transition Plan). Following on from this, a second phase of research identified niche, near-to-market, or prototypical tools (e.g. material devices or ICT supports), rules (e.g. regulations or targets) and skills (e.g. information, learning opportunities) that mapped onto the envisaged short-term interventions. With the research team acting as explicit intermediaries in the process, a range of households in Dublin were invited to participate in the testing and evaluation of these interventions in a series of HomeLabs.

Initially focusing on how the visions, institutions, intermediaries and technologies of the HomeLabs were assembled and implemented, the remainder of this chapter outlines the outcomes of the Washing HomeLabs. Specifically, the experiences of participating households are counterposed against the public reactions, both in the mass media and on the streets, to the significant public sector restructuring in the water sector.

3. Assembling HomeLabs

As indicated by the wealth of examples documented in this volume, ULL can assume many forms and take place in diverse settings. Nonetheless, commonalities across these applications can be discerned in terms of widespread commitments to co-creation, exploration, experimentation and evaluation. This phased process was a key feature of the CONSENSUS project culminating in the development, implementation and review of the HomeLabs experiments (Davies, 2014; Davies et al., 2012; Devaney and Davies, 2016).

The CONSENSUS approach enabled a range of stakeholders, many of whom were rarely brought together in their professional settings, to collaboratively imagine and co-create visions of more sustainable washing futures. The resulting

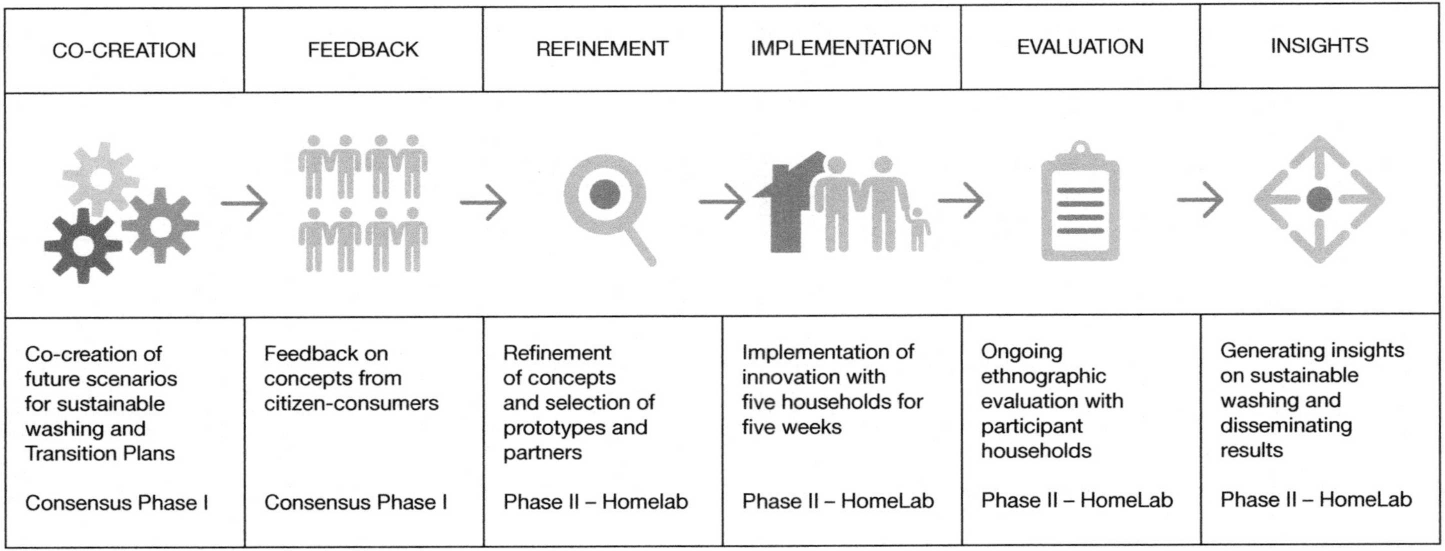

CO-CREATION	FEEDBACK	REFINEMENT	IMPLEMENTATION	EVALUATION	INSIGHTS
Co-creation of future scenarios for sustainable washing and Transition Plans	Feedback on concepts from citizen-consumers	Refinement of concepts and selection of prototypes and partners	Implementation of innovation with five households for five weeks	Ongoing ethnographic evaluation with participant households	Generating insights on sustainable washing and disseminating results
Consensus Phase I	Consensus Phase I	Phase II – Homelab	Phase II – HomeLab	Phase II – HomeLab	Phase II – HomeLab

FIGURE 8.1 CONSENSUS living laboratory approach

scenarios articulate a combination of lifestyle, technological and governance innovations that might combine to support more sustainable washing practices. Following contributions from citizens brought together through a series of workshops, the concepts perceived to be the most promising and desirable were refined and devices, services, ideas and simulated regulatory conditions that mirrored those ideas were identified. Three promising practices were identified: connections to nature, adaptive washing and efficiency in use and a number of regulatory, norm disrupting and technological interventions identified for each (see Table 8.1), including a shower litre metre, low flow showerhead devices, cleansing products that reduced the need for water use (e.g. two-in-one shampoo and conditioner), communication channels to exchange norm disrupting ideas, messages, discussions (provided through various social media channels identified by the participants as relevant to them). This process involved extensive inter-action with innovators, designers, producers, retailers, researchers as well as policy makers and public utilities to identify suitably disruptive interventions that could be easily incorporated into the fabric of the participating households lived environment.

Interventions were introduced to households over a five-week period on a step-wise basis. A mixed method, ethnographic approach was then developed to facilitate the review and evaluation of the interventions. This process also enabled participants to reflect on their experiences of participating in the HomeLabs process itself. During the HomeLabs, each household was visited by a researcher once a week to explore experiences, collect data and brief participants on forthcoming interventions. Semi-structured interviews were utilised as the primary form of data collection during these visits, supplemented by photographic evidence to capture physical contexts and material environments. Social media was used to engage with participants throughout the HomeLabs with prompts, texts, Facebook posts and WhatsApp interactions employed where appropriate to stimulate discussion and prompt active engagement with the supporting interventions provided to households.

In terms of recruiting households to participate in the HomeLabs, the aim was to involve people inhabiting different household structures and life stages across diverse socio-economic demographics, which might illustrate contrasting and contextualised challenges, opportunities and experiences. It was not to identify a representative sample of households in any statistical sense. As a result, households with the most common structures in Ireland, as identified by the Central Statistics Office (2012), were invited to participate through group mail outs, online advertisements, snowballing and flyers at events held in Dublin over the summer of 2014. Across the HomeLabs, over 40 households in the Greater Dublin Area and commuter suburbs matching the desired criteria volunteered as possible participants. This number was reduced to ten following attention to practical issues such as householder availability, accessibility and showering infrastructures, with five households ultimately participating in the Washing HomeLabs (Table 8.2).

TABLE 8.1 Washing HomeLabs research process

HomeLab framework	Week 1: Baseline	Week 2: Connected	Week 3: Efficient	Week 4: Adaptive	Week 5: Wrap-up
Practice dimension	Establish current habits and practices	Enhance understanding of water services and water availability	Identification of litre targets and prompting lower flow and social feedback	Trial less familiar practices that enable more substantial water reduction	Users adopt preferred practices based on experience and impacts
Governance: Rules and regulations Targets for average water consumption in litres per person per day	No targets	40 litres	25 litres	15 litres	No targets
Tools: Devices to enable participants to measure and manage consumption Hair and personal care products that may facilitate reduced water use	Shower litre meter Shower timer Users existing products	Shower litre meter Shower timer Low foam shampoo Leave-in conditioner Hair and body wash	Shower litre meter Shower timer Low flow showerhead 2-in-1 shampoo and conditioner Co-wash	Shower litre meter Shower timer Low flow showerhead Dry bath product Dry shampoo product	Shower litre meter Shower timer Low flow showerhead User selected products
Skills and understandings: Behavioural guidance and motivational information		Communications on water cycle	Shower pausing and flow adjustments Costs and comparisons	Reduced washing and splash-washing Norms challenging literature	
Research Process: Methods used to gather data	Shower logs WhatsApp Home visit	Shower logs WhatsApp Home visit	Shower logs WhatsApp Home visit	Shower logs WhatsApp Home visit	Shower logs WhatsApp Home visit

TABLE 8.2 Households recruited in the washing HomeLabs

Profile	Couple (C) Household	Family (F) with **Young** (Y) **Children**	Mixed (M) Household (i.e. **Non-Familial**)	Family (F) with Teenagers (T)	Family (F) with Adults (A)
Identifier[1]	Household C	Household FY	Household M	Household FT	Household FA
Occupants	2	5	4	4	4
Name[2] (and Age)	Amy (29) Darren (32)	Sam (42) Laura (40) Edel (13) Connor (9) Jack (7)	Damian (35) Ruth (36) Martin (36) Alison (33)	Gareth (50) Kathy (49) Ronan (18) Jill (16)	Aisling (61) James (63) Peter (25) Claire (21)
Home type	Apartment	Bungalow	Terraced flat	Semi-detached	Semi-detached
Ownership	Owner-occupied	Owner-occupied	Rented	Owner-occupied	Owner-occupied
Shower flow (litres per minute)	13.8 LPM	13 LPM	7 LPM	8 LPM	7.5 LPM

1 The Household keys outlined here (Household C, FY, M, FT and FA) are used as identifiers in the results section along with the name and ages of participants.

2 These are pseudonyms to protect the identity of participants.

The practice of personal washing was selected for analysis because while water resources remain relatively abundant across much of Ireland currently, climate models are predicting increased water stress in the future especially in urban areas. At the same time, water consumption per capita per day in Ireland is estimated to be one of the highest in Europe with demand predicted to expand due to population growth and increasing standards of living, placing increased stress on the antiquated systems of water supply and treatment (Li et al., 2010). In reality, data on actual water use is limited and even less is known about the hyper-privatised world behind the shower curtains across Ireland (Davies and Doyle, 2015b).

4. Washing HomeLabs and water governance in Ireland

The development of the HomeLabs took place during a period of extreme flux across much of Irish society. Following the economic crash in 2008, employment rates plummeted in many sectors and public sector budgets were cut drastically, reducing salaries and service provision across the board, undermining strategic plans for infrastructural development (Kearns et al., 2014). Every opportunity to cut costs or raise revenue was debated within government and between the government and the Troika, who were overseeing the financial bailout provided to prevent the bankruptcy of the Irish State. Up until this point, the Irish public had largely avoided direct metering and water pricing typically associated with the neoliberalisation of water governance elsewhere (Bakker, 2005; Castree, 2008). Indeed, Ireland was the "only country in the OECD where households [. . .] [did] not pay directly for the water they use[d]" (Bord Gáis, 2013, p. 7), instead the costs were covered through the general taxation system. In this context, it was inevitable that the cost of providing water services would come under scrutiny. There was little surprise when in 2009 the Minister of Finance announced that preparations for nationwide metering and water charges were underway. What ensued was a political tug of war between the coalition governing partners (the Labour Party and Fine Gael) and opposition parties. The much vaunted willingness of the Irish populace to accept cuts in pay and services in an attempt to balance the books following collapse of the banking sector (McMahon, 2015), was however, severely tested as a result. In the light of widespread pay cuts, additional taxes and reduced public services, the spectre of water charges became the final straw for many households struggling to cope under austerity and a series of mass rallies and protests ensued (Hearne, 2015).

Such was the extent of water sector reform that it has been characterised as one of the most ambitious remodelling programmes ever undertaken by the Irish State (Bord Gáis, 2013). The restructuring process was not straightforward however, as indicated by the timeline of water reform detailed in Table 8.3. Supported by the Water Service Act of 2013, charges were initially to be based on volumetric consumption, calculated by metres that Irish Water would install in each household. However, the development of the new system has been characterised by delays, revisions and uncertainties. Alongside the programme for metering and pricing came the formation of a semi-state national water utility, Uisce Éireann (Irish Water), to replace the previous 34 local water authorities and provide a mechanism through which the self-financing of water provision might take place, regulated by the Commission for Energy Regulation (CER). Justified as a necessary mechanism to respond to the ailing water infrastructure, which had been starved of investment for decades, and the fiscal constraints of the Irish State, the new organisational and technical composition of water governance saw the extension of financial logics and a highly technical application of environmental metrics into the water realm in Ireland for the first time; a process that has been called bio-financialisation (Bresnihan,2016).

While revenue raising was clearly a primary driver for reform of the water sector, arguments that the introduction of metering would lead to greater efficiency through improved data collection on usage and enhanced ability to detect leaks were also emphasised by governing actors. Irish Water frequently cited international research which found that installing metres led to reduced water consumption, at least in the short-term, of around 20 per cent (OECD, 2010). However, such claims were refuted by oppositional actors who pointed out that the metres to be

TABLE 8.3 Irish water milestones

Year	Month	Event
2009	December	Preparations for water charges announced with fees for volumetric consumption above a fixed volume to be provided for free
2010	June	Labour leader (in coalition with Fine Gael) rules out water charges on the basis of the cost of installing metres
	November	National Recovery Plan rules out interim flat rate of residential water tax pending installation of meters in 1.2 million homes. Memorandum of understanding on the Troika bailout states that water charges will be introduced in 2012 or 2013, when metres have been installed across the State. New national water utility Irish Water announced.
2012	April	Government confirms water charges will finance the estimated €800 million needed to reform the water sector. Announcement that the contract to run Irish Water was awarded to Bord Gáis Éireann, a national energy utility. Announcement that charges will be introduced in 2014
2013	July	Irish Water incorporated as a semi-state company under the Water Service Act 2013
	December	Water Service (No. 2) Bill 2013 progresses through all stages in the Dáil over four hours amid protest from the Opposition
2014	January	€85 million consultants bill announced by Irish Water
	May	Government agrees to eliminate €50 standing charge within water charges
	July	CER provides details of proposed water charges indicating that the average cost for a household of two adults and two children will be €278 with metered rates set at €4.88 per 1,000 litres
	October	CER grants one month extension to registration period for Irish Water following low registration rates. Opposition to the charges intensifies as an estimated 50,000 protest in Dublin. The Green Party calls for constitutional guarantees that water services will remain publicly owned
	November	Revised Water Services Act 2014 including a cap on annual charges; a water conservation grant €100; PPS no longer required for registration with Irish Water; statutory dispute resolution system; abolition of allowances replaced with cap on charges and reduced water rate from €4.88 to €3.70/1,000 litres; establishment of public water forum; capped charges until December 2018
2015	April	First bills issued (covering the period January–March 2015)
2016	January	Irish Water announce that 928,000 customers (or 61% of their customer base) had paid their bill in at least one of the three cycles up to that point
	April	A new Fine Gael-led minority government is formed 63 days after the general election with the future of Irish Water and water charges a key negotiating point between the leading parties. Agreement was established that water charges would be suspended for nine months, while a new "commission" prepares a report for the Dáil on how Irish people should pay for water in the future

rolled out across Ireland were not state-of-the-art in terms of their functionality and would not be user-facing in terms of their output. Indeed, the metres were installed underground, under manhole covers outside household footprints on the public thoroughfare. The digital metres, fitted with tamper-proof technology, were designed to be read remotely by Irish Water for billing purposes with monthly usage details being noted on customer bills. People are actively discouraged from attempting to read their metre with the Irish Water website stating, "[d]o not try to read the water metre yourself as it may involve bending, crouching and kneeling on the ground" (Irish Water website, Accessing your water meter, 2016). As such, the metering and its data provide little assistance to householders with an interest in monitoring and managing their consumption of water in the home in real-time. This disconnect between charging and data availability further compounded the view that the reform of the sector was simply a revenue raising exercise being initiated to bolster liquidity as a precursor to eventual privatisation of the services. Certainly, little emerged from Irish Water in relation to sustainable consumption or domestic water conservation during its early years of operation and when information did emerge in 2014, it was highly simplistic and in stark contrast to the nuanced approach enshrined within the Washing HomeLabs.

The Washing HomeLabs were developed at precisely the time that the water sector was in this period of highly contested flux. This was initially perceived by the research team as a fortuitous opportunity to experiment with and test the resulting outputs from multi-stakeholder brainstorming about sustainable futures to inform the very architecture of the emerging institutional infrastructure for modern water services within Ireland. Indeed, Irish Water was invited to participate in the Advisory Group for the HomeLabs and a policy specialist was identified to attend meetings and engage with inputs. However, on-going negative publicity around Irish Water, not least in relation to the scale of consultant's costs in the establishment of the organisation, uncertainty over charges, the regular public protests in Dublin and around the country during installation of metres meant that in reality there was little dialogue between the body and the research team during the HomeLabs experiments. In particular, requests to identify metred households for the experiments that would secure access to accurate overall water usage data over the duration of the HomeLabs were rejected. A reflection on this and subsequent interactions follows the brief discussion of the HomeLabs process and findings below.

4.1. Washing HomeLabs

Once the suite of interventions had been identified with the stakeholders, a phased plan for their combination and alignment was drawn up (see Table 8.1). The aim here was to script and disrupt practices of washing within households in order that participants washing practices became more ecologically connected with their water supply, more efficient in their use of water and more adaptable to water availability. Overall, integrating the interventions relating to governance (e.g. water targets), tools (e.g. low-flow shower heads) and education (e.g. information provided through the water portal) yielded positive changes in washing practices across all participants across the initial five-week experiment. Indeed, an average reduction of 47 per cent in water use per person per day for personal washing was achieved, although this masks a wide variation in practices among individuals within households. Heterogeneity existed in washing practices both within and between households of different types. Overall, six key types of washing practice were identified with people tending to adopt between one and three of the key types of washing practice over the research period (Table 8.4) and opportunities to target interventions according to these were identified. The resonance of these washing types among the wider population was interrogated through the WASHLab exhibit, which formed part of the HOME\SICK exhibition at the Science Gallery in Dublin in 2015 (Davies et al., 2015). More than 2,000 visitors to the exhibition engaged with the exhibit, which included an interactive installation and tailored digital animation shorts providing information on how to reduce water consumption for personal washing. After the intensive five-week HomeLabs experiment, the participating households were left with all the devices, tools and norm disrupting supports. All households agreed to participate in follow-up surveys to be completed after six months and twelve months.

Full details of the Washing HomeLabs, their structure, function and performance are documented elsewhere (Davies et al., 2015; Davies and Doyle, 2015a, 2015b), but in terms of interpreting the key findings, it is important to reiterate that all participants had explicitly agreed to engage in the HomeLabs experiment. Nonetheless, during discussions participants also suggested that their commitment to embedding such endeavours in their everyday lives beyond the experimental framework would depend on them receiving convincing evidence from trusted experts of the need to reduce water use. Certainly the participants were yet to be convinced around the need to reduce water use in Ireland particularly in relation to matters of water scarcity, even though this is an area that climate change modelling already suggests this will increasingly be a concern for Ireland in the future (Sweeney, 2008). Similarly, participants were not clear exactly why they should be washing their hair and bodies from a hygiene perspective. This is in stark contrast to hand washing where the need to wash to prevent the spread of diseases was widely appreciated. So certain conditions were yet to be met, as far as the participants were concerned, around the *need* to take steps to reduce water use for washing purposes. It was also felt that there were limits to the amount by which water use for washing could be reduced given current social expectations, infrastructures of water provision, the physical architecture of showering and bathrooms and the availability and affordability of supporting products and devices.

In terms of participant feedback on specific interventions, the Water Portal – a digital platform that allowed researchers to provide information, prompts, data and a space for dialogue about the challenges and benefits of different supports and their interaction – was considered an important motivator for households containing information on water sources, availability, quality, pricing,

TABLE 8.4 HomeLabs washing types

Washing type		Description
	The routine refresh	Showers where the aim is primarily to refresh the physical appearance and to feel invigorated; often involving a hair wash
	Post-exercise clean	Showering prompted by exercise and associated feelings of uncleanliness where freshening up is key
	The wake-up shower	Highly routinised and engrained morning showers where wake-up and psychological functions of "starting the day" are primary motivations, in addition to feelings of cleanliness after a night's sleep and physical presentation for work
	Intensive grooming	Showers that go beyond basic maintenance and include hair treatments, shaving, body scrubs, facials; often in preparation for going out
	Therapeutic	Showers or baths performed for therapeutic reasons where hot water, muscle relief, relaxation and recovery play key roles
	The escapist shower	Showering in excess of completing cleanliness functions where the shower is seen as a refuge for zoning out and escaping the pressures of daily life

and consumption targets. However, it was also clear that while participants were happy to "play along" with the HomeLabs experiment over its delineated time period, this participation was reliant on the no-cost implications to household budgets (all devices, tools, products and supports were provided free of charge). Participants were adamant that they would need a more robust justification to undertake further investment (financial or social) to reduce their water use in relation to personal washing. Equally, it was felt that any water consumption targets for personal washing would require positive relations of trust between providers, regulators and consumers (or customers as Irish Water refers to them) that it was felt had not developed between Irish Water and the public at the time of the experiment.

In contrast to the widely documented unrest among large swathes of the Irish public around metering (see Hearne, 2015), many participants found the user-facing devices to indicate volumetric water consumption particularly useful in reconfiguring their washing practices. Indeed, prior to the HomeLabs there was widespread underestimation of water used for washing across all participants, so the visualisation of water consumed in precise washing events was found to be informative, providing a necessary baseline from which changes could be measured. It also revealed the diversity of what goes on behind the shower curtain in ways that simple timing of washing events fail to do. Practices of pausing or reducing flow during phases of washing or particular grooming activities varied widely, as did the extent and regularity of certain activities, such as hair washing or shaving, within the shower. This real-time visualisation of water consumption provided during the HomeLabs contrasts starkly with the information flow from the digital metres installed by Irish Water, which were deliberately located not just outside the bathroom, utility room or kitchen, but outside the physical footprint of the houses consuming the water. The burying of the metres beneath layers of "protective" heavy metal covers and their location on public property (pavements and roads), works directly against any responsive monitoring of metering information by households.

Equally, there are contradictory messages from Irish Water with regards engagement with the water metres. As detailed earlier, there are explicit messages suggesting that householders should not attempt to read their metres themselves (instead relying on the data that are presented in quarterly bills), yet in other sections of their website they claim that the water metres will allow "you [the consumer] to understand your household's water usage and minimise your impact on the environment" (Irish Water website, Water Conservation, 2016). How, exactly, such understanding and minimisation is to take place is not detailed, but the insights provided for householders by Irish Water in relation to water conservation clearly adhere to the view that consumption behaviour is driven primarily by an information deficit. This is most clear in their headline statement that, "[b]y being more aware of conserving water in our homes, we can contribute to building a sustainable water supply for everyone" (Irish Water website, Water Conservation, 2016). There is no mention here of the infrastructure needed for conservation (whether that be in-house or in terms of the reduction in leakage across the water supply system) or societal norms and habits that set out everyday expectations around washing. Certainly, the broad statements provided do not provide any of the kinds of information that HomeLabs participants said they would need to convince them that water reduction is required. Indeed, the HomeLabs indicated that for some households with power showers and a predilection for long therapeutic washing practices, having a bath would actually use less water. Similarly, the HomeLabs indicated that because of the variegated washing practices within the shower (through pausing or reducing flow while lathering, for example), the correlation between time and water consumed was certainly not as linear as the Irish Water illustration suggests. It is also interesting that Irish Water talks specifically about saving energy, water and money through water conservation techniques, yet there is no attempt to quantify the impacts that the water conservation tips they provide might make cumulatively make to households across these arenas.

Despite the potential for remote reading of the metres by Irish Water employees in formulating bills, no such remote reading capabilities are provided for those consuming the water. The absence of user-facing data of water use, outside the cubic metrage consumed (which is detailed on quarterly bills for metred properties), dramatically reduces the potential of nationwide metering to provide meaningful information for reducing domestic water consumption. However, the HomeLabs indicated that opportunities to reduce water use through combining the visualisation of water consumed at point of use with innovation in both shower cubicle and shower unit design (e.g. to provide for efficient access to cleansing products and responsive settings for showers) and innovative hair and body care products (e.g. removing cleansing steps such as conditioning hairs from the shower) are manifold. These opportunities for efficiencies were, though, contingent upon the experiences and outcomes of innovations meeting the expectations of users in terms of performance. Low-foaming or no-rinse hair and body products needed to smell nice have a good texture and leave skin or hairs in the desired condition to be used; water saving capacities were alone insufficient to convince users to ditch their tried and tested product range.

Nonetheless, beyond the regulatory simulations and the socio-technical developments in devices and products, participants were often surprised just how easy they found incorporating some "disruptive" ideas into their washing routines. Such disruption was achieved by encouragement to experiment with enhanced planning of showering events to accommodate social or exercise schedules or prompting participants to consider alternative, less resource intensive, ways to achieve end goals such as revitalisation or relaxation. While generally received positively, these disruptive ideas were met with foundational concerns regarding health and hygiene. With the exception of handwashing, participants felt there was a lack of guidelines regarding how to wash and the frequency and duration of that washing in order to achieve appropriate hygiene standards. In the absence of such information, certain participants involved in the HomeLabs provided interesting insights into the actors or information sources participants responded to for the refinement of their social norms around washing. These actors included on-line fashion, fitness or beauty bloggers and hair stylists who were felt to offer authoritative advice on products and lifestyles to achieve a "fashionable" appearance. Of course, there are manifold challenges in attempting to enrol such distributed and diverse intermediaries in any form of collective endeavours around sustainable washing, but it is clear that mainstream mass media advertising and marketing is no longer the sole, or indeed even the main, source of information for establishing social norms.

Overall, the high-level findings of the HomeLabs indicate significant merit in aligning and combining interventions, which support and facilitate more sustainable household washing practices. In particular, the central role of the researcher in the efficacy of the experiment, providing on-going motivation, information, interpretation and explanation, was clear. This intermediary role is something that is increasingly being discussed in the realm of sustainability transitions at a variety of scales (Guy et al., 2011). Certainly, during the intensive research period, researchers acted as crucial intermediaries, sourcing and installing products and devices and providing access to information and training. As a result, households were able to easily disrupt their habitual behaviour and build new skills and understanding around their consumption practices, at least in the short-term. However, the HomeLabs also sought to establish whether participating in the experiment had any long-term impact on washing practices and a survey was completed by participants prior to the HomeLabs study in 2014, and then again after six and 12 months, to explore the extent to which any changes induced by the study were maintained. The survey explored the washing activities performed and, drawing on Verplanken and Orbell's (2003) Self-Reported Habit Strength Index, examined the frequency and automaticity of their washing practices.

Nearly a quarter of all participants in the Washing HomeLabs reported an increase in the sustainability of their washing practices one year on from participating in the HomeLabs, with showering and bathing frequency declining on average across all households. In terms of the strength of habitual washing practices, efforts to take a shorter shower also remained a strong habit for households twelve months on as did turning off the tap while brushing teeth. Indeed, the strength of the habit had increased across each household bar one. The use of the shower timer and/or the litre metre was adopted by participants in four out of five households and the participants felt that the litre metres provided a useful visual cue of water use. They also felt though that there were currently few incentives to motivate people to purchase and install such metres given the current capping on water charges.

In early 2016, water charges were a maximum of €160 per year for single-adult households, with a household of two adults or more having their bill capped at €260 per annum. However, debates around water charges played a significant role in political campaigns during the 2016 General Election and they were eventually suspended in March of that year pending a report of an appointed Expert Water Commission. The political landscape around the governance of water remains tense and variegated. In 2016, Fine Gael, the ruling minority government, supported retaining water charges and Irish Water with a view to extending the cap on charges beyond 2018 until 2021, as did Labour albeit alongside a universal free basic allowance. Fianna Fáil (the next largest political party in terms of elected officials) favoured a slimmed-down utility in place of Irish Water. In contrast, Sinn Féin supports the abolition of charges and Irish Water with refunds for citizens who have already paid water charges. Both AAA-PBP (Anti-Austerity Alliance-People Before Profit) and the Social Democrats seek the abolition of charges and for Irish Water to be scrapped, while the Green Party favours a publicly owned water service with charges, but with a basic free allowance for all.

In November 2016, an Expert Water Commission, set up in the light of the intense interactions over the future of water charges and Irish Water before and during the election, reported its initial recommendations on funding public domestic water services in Ireland. The Commission, composed of international and national actors involved in water governance, including the OECD's environmental director, recommended that funding of water services, for normal domestic and personal use should be made as a charge against taxation. It goes on to recommend that "[t]he volume of water necessary to meet the *normal* domestic and personal *needs* of citizens should be independently assessed through an open and transparent process" (Expert Water Commission, 2016, p. 1, emphasis added), while "*[e]xcessive* or wasteful use of water will be discouraged by charging for such use" (Expert Water Commission, 2016). Recognising the difficulties in identifying "normal" usage, the Commission suggests calculating normal based on current actual consumption of water, but placing limits on the boundaries of acceptable use. For example, levels could be set that correspond to the actual consumption of a significant proportion of water users (e.g. 90 per cent of users) or at a particular level above the current average domestic consumption (e.g. at 150 per cent of the average). Alternatively, the Commission suggests establishing "normal" by calculating volumes based on standard uses for domestic water consumption such as personal washing, toilet flushing, drinking, cooking, clothes washing, dishwashing, waste disposal and house cleaning, but this would require much more detailed analysis to establish the precise levels of allowance to be made available. It also assumes that current or "standard" usage is sustainable and does not distinguish between "needs" and "wants".

In line with recommendations made by the CONSENSUS team in their Transition Framework for sustainable washing (Doyle and Davies, 2012) and the HomeLabs high level findings report (Davies and Doyle, 2015a), the Commission also suggests that the consumer's voice should be at the heart of discussion and decision-making on the delivery of water services in Ireland. Recognising that insufficient attention has been paid to social governance and the engagement of civil society, the Commission also calls for a considerably more proactive approach to be adopted in relation to promoting domestic water conservation measures. It is acknowledged that this engagement must go beyond educational and information campaigns, to include advice and access to water conserving devices and modifications to existing building regulations to require the incorporation of water conserving fittings in new buildings. While commendably extending current debates on water services, there is still a technological and efficiency narrative which dominates the Commission's findings. Nowhere, for example is there a consideration of, or even a call for more attention to, the social norms, for example around personal washing practices, that impact on water use and which have been shown in the HomeLabs to be highly differentiated both across and within households of different structures and composition. Similarly, the call for an EPA administered research budget on water management and conservation is welcome, but singularly fails to acknowledge that the EPA have already funded research in

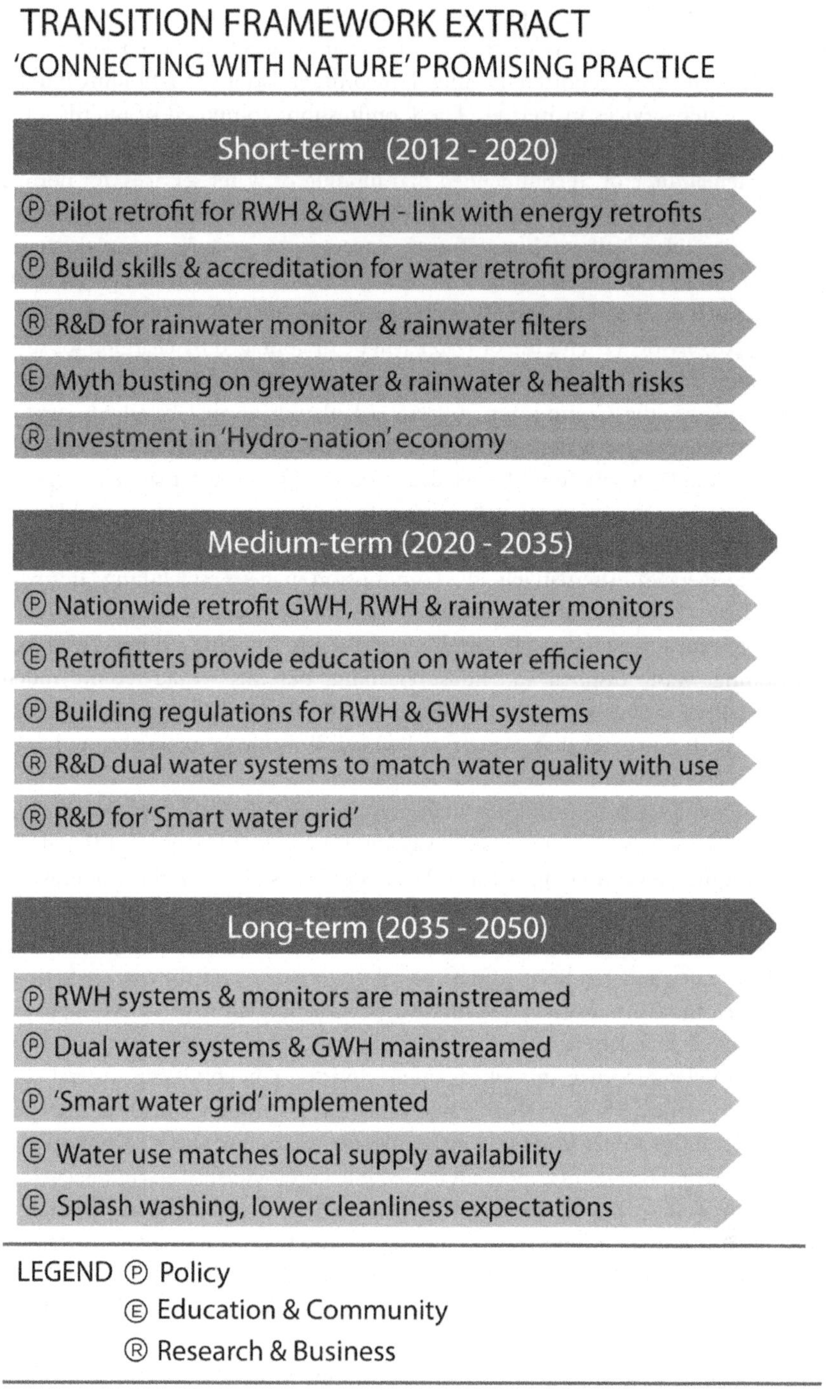

TRANSITION FRAMEWORK EXTRACT
'CONNECTING WITH NATURE' PROMISING PRACTICE

Short-term (2012 - 2020)

Ⓟ Pilot retrofit for RWH & GWH - link with energy retrofits

Ⓟ Build skills & accreditation for water retrofit programmes

Ⓡ R&D for rainwater monitor & rainwater filters

Ⓔ Myth busting on greywater & rainwater & health risks

Ⓡ Investment in 'Hydro-nation' economy

Medium-term (2020 - 2035)

Ⓟ Nationwide retrofit GWH, RWH & rainwater monitors

Ⓔ Retrofitters provide education on water efficiency

Ⓟ Building regulations for RWH & GWH systems

Ⓡ R&D dual water systems to match water quality with use

Ⓡ R&D for 'Smart water grid'

Long-term (2035 - 2050)

Ⓟ RWH systems & monitors are mainstreamed

Ⓟ Dual water systems & GWH mainstreamed

Ⓟ 'Smart water grid' implemented

Ⓔ Water use matches local supply availability

Ⓔ Splash washing, lower cleanliness expectations

LEGEND Ⓟ Policy
Ⓔ Education & Community
Ⓡ Research & Business

FIGURE 8.2 Washing transition framework

this field which was not incorporated into Irish Water's work programme. Attention to how future research around water management and conservation is framed and how findings are to be received and considered by Irish Water will be crucial to the efficacy of any future funding schemes.

5. Conclusion

The HomeLabs offered a protected space to trial novel, niche or prototypical policy, product and behavioural interventions for more sustainable washing practices. The "lab" method and concept was particularly useful as it explicitly elevates the participants (the households in this research) to become research partners rather than subjects and the practice focus enabled the research to go beyond technology- or regulatory-driven behaviour change in the home. Equally, technologies were considered for their transformative potential, as "assistors" in supporting more sustainable washing practices, not just in terms of their functional capabilities. While the overall structure of the research was common across households the high levels of participant input helped to generate creative, impactful and empathetic interventions more suited to the participants needs, contexts and experiences.

That the HomeLabs were spatially and temporally delimited experiments with agreed parametres and roles was both a benefit and a challenge. Participants were not committing themselves to any long-term financial or behavioural changes, but were explicitly permitted through the "lab" framing to be experimental; to do things differently without facing negative peer-pressure. Participants became the "lead scientists" in testing products, devices and novel norm disrupting behaviours in their everyday environments, feeding their experiences back to trusted and responsive researchers acting as key intermediaries between them and other stakeholders seeking to develop novel solutions, goods, services and business models. The HomeLabs represented a space to debate and test sustainability transitions with those often called on to "take action" to protect the environment and use resources more efficiently. As with broader living laboratory methodologies discussed elsewhere in this book, the feedback and reactions obtained from householders provided a foundation for the iteration and improvement of the supplied tools, rules and educational interventions aimed at mainstreaming supports for more sustainable domestic washing. Reflecting on the results provided in this chapter, the HomeLabs methodology also delivered additional benefits in terms of provoking deliberation, discussion and experimentation among participants in relation to their everyday washing practices, albeit in many different ways depending on those involved. The need to consider these differentiated capabilities, reactions, preferences and priorities when it comes to achieving future behaviour change is crucial and the HomeLabs directly addressed this need. Essentially, they progressed ideas about future sustainable consumption from abstract scenarios to concrete realities, highlighting the benefits of coordinating technological, policy and informational supports with appropriate motivating forces through governing arrangements for optimising behaviour change.

As detailed in Davies and Doyle (2015a), a challenge for experiments such as HomeLabs, is how positive outcomes and learning identified in bounded experimental sites might be rolled out. The intensity of human resources involved in establishing, running and evaluating HomeLabs-style initiatives means a simple "scaling-up" of such activities nationwide is unrealistic. International research on household water practices suggests that the commonly articulated policy goal of "scaling-up" in the water sector may itself be inappropriate, ignoring the contingent complexities and entanglements of society, culture and politics (Fam et al., 2015). What the HomeLabs do indicate is that tailoring supports in a way that resonates with people's needs and accommodates their physical and social context is more likely to optimise the extent to which change will both occur and become habituated. Equally, while technological fixes will inevitably form an important part of any shift toward more sustainable consumption, the socio-political dimension of water use and washing must not be underestimated.

The creation of Irish Water has, without question, entirely rescaled water governance in Ireland. With the centralisation of water supply and treatment and the introduction of charges, water governance in Ireland is transcending its previous sub-national and rather technical configuration, moving directly into the highly contested terrain of financialisation and commodification. The progress of these governance changes has been marked by a series of announcements, outcry and revision. Unfortunately, the highly charged atmosphere within which the reform has taken place has left little space for articulating arguments around sustainability transitions. There is scant evidence that the new regime seeks to develop expanded relationships with citizens around the sustainability of their use of water beyond that of a "customer". As their website states,

> we'll need you to work with us to ensure we have sustainable water services into the future. To do this we need your participation when it comes to metering, registration, and eventually billing. Together we can improve and secure this precious resource that will be vital to the social and economic life of this nation far into the future.
>
> (Irish Water, Why value water?, 2016)

Certainly, there is nothing to suggest that Irish Water has any interest in deepening its understanding of what goes on "behind the shower curtain". Clearly based on a predict and provide approach, Irish Water is focused primarily on the technical solutions required to meet infrastructural weaknesses in the water supply system so that "Ireland has a water network that can accommodate all our needs in the near future" (Irish Water website, Why value water?, 2016).

In conclusion, while reform of the Irish water sector was long overdue, the changes made thus far have made little progress in terms of addressing many social, economic and environmental questions of sustainable water supply and treatment. In the absence of direct and explicit strategic niche management for sustainability transitions from the Irish government, experimentation through HomeLabs provided a less directed process, with diverse actors leading actions to stimulate new spaces of innovation for the testing and evaluation of alternative washing practices. A benefit of such organisation is autonomy; the HomeLabs were designed, operated and analysed without direction from political actors or public institutions. This was fortuitous given the fracturing of trust between Irish Water and many sections of Irish society that occurred over the duration of the research. The independence of the HomeLabs certainly aided the decision of participants to give their time freely, to allow the novel interventions in their hyper-privatised world of washing and, most importantly for the research, to engage actively in testing and evaluation. However, such autonomy also potentially limits the impact of the findings as links to actors with policy making powers are less well established and sustainability agendas can remain side-lined if they do not fit in to existing political landscapes. Certainly, sustainability with regards to water governance in Ireland has been primarily debated among politicians with respect to matters of financialisation, with water charges forming part of a suite of austerity measures following a deep and, for much of the Irish population, painful recession. In this context, attention to environmental and social sustainability concerns remains steadfastly subservient to economics.

References

Bakker, K. (2005) 'Neoliberalizing nature? Market environmentalism in water supply in England and Wales'. *Annals of the Association of American Geographers*, *95*(3), 542–565.

Bord Gais. 2013. 'Transforming water services in Ireland'. accessed on 10 March 2014 Retrieved from http://westmeathcoco.ie/en/media/Irish%20Water%20Leaflet.pdf (accessed on 10 March 2014).

Bresnihan, P. (2015). 'The bio-financialization of Irish water: New advances in the neoliberalization of vital services'. *Utilities Policy*. doi:10.1016/j.jup.2015.11.006

Bulkeley H. and Castán Broto V. (2012). 'Government by experiment? Global cities and the governing of climate change'. *Transactions of the Institute of British Geographers*, *38*, 361–375.

Camaren, P. and Swilling, M. (2014). 'Linking complexity and sustainability theories: Implications for modeling sustainability transitions'. *Sustainability*, *6*, 1594–1622.

Castree, N. (2008) 'Neoliberalising nature'. *Environment and Planning*, *40*, 131–152.

Davies, A.R. (2014). 'Co-creating sustainable eating futures: Technology, ICT and citizen–consumer ambivalence'. *Futures*, *62*(B), 181–193.

Davies A.R. and Doyle R. (2015a) 'Transforming household consumption: From backcasting to HomeLabs experiments'. *Annals of the Association of American Geographers*, *105*(2), 425–436.

Davies, A.R. and Doyle, R. (2015b) 'Waterwise: Extending civic engagements for co-creating more sustainable washing futures'. *ACME: An International E-journal for Critical Geographies*, *14*(2), 390–400.

Davies, A.R., Doyle, R. and Pape, J. (2012). 'Future visioning for sustainable household practices: Spaces for sustainability learning?' *Area*, *44*(1), 54–60.

Davies, A.R., Fahy, F. and Rau, H. (2014). *Challenging consumption: Pathways to a more sustainable future*. London: Routledge.

Davies, A. R., Fahy, F., Rau. H. and Pape, J. (2010). 'Sustainable consumption: Practices and governance'. *Irish Geography*, *43*(1), 59–79.

Davies, A. R., Lavelle, M.J. and Doyle, R. (2015). 'WashLab summary analysis, Trinity College, Dublin'. Retrieved from http://consensus.ie/wp/sample-page/papers-reports/.

Devaney, L. and Davies, A.R. (2016). 'Disrupting household food consumption through experimental HomeLabs: Outcomes, connections, contexts'. *Journal of Consumer*, online advanced access. doi: 10.1177/1469540516631153

EPA. (2006). 'Emissions inventory for Ireland 2006'. London: EPA.

EPA. (2012). 'Emissions inventory for Ireland 2012'. London: EPA.

Fam, D., Lahiri-Dutt, L. and Sofoulis, Z. (2015). 'Scaling down: Researching household water practices'. *ACME: An International E-journal for Critical Geographies*, *14*(3), 639–651.

Guy, S., Marvin, S., Medd, W. and Moss, T. (2011). *Shaping urban infrastructures: Intermediaries and the governance of socio-technical networks*. London: Routledge.

Hearne, R. (2015). *The Irish water war, austerity and the 'Risen people': An analysis of participant opinions, social and political impacts and transformative potential of the Irish anti water-charges movement*. Maynooth University. Retrieved from https://maynoothuniversity.ie/sites/default/files/assets/document/TheIrishWaterwar_0.pdf (accessed 16 February 2016).

Kearns, G., Meredith, D. and Morrissey, J. (2014). *Spatial justice and the Irish crisis*. Dublin: RIA.

Kelly, F. (2016). 'Water charges cap may be extended until 2021'. *Irish Times, 16 January 2016*, online edition retrieved from http://irishtimes.com/news/politics/water-charges-cap-may-be-extended-until-2021-1.2498830 (accessed 18 January 2016).

Lavelle, M. J. (2013). *Towards sustainable consumption: An empirical investigation of attitudes and behaviours towards household consumption and sustainable lifestyles in Northern Ireland and the Republic of Ireland*. Unpublished PhD thesis, National University of Ireland, Galway.

Li, Z., Boyle, F. and Reynolds, A. (2010). 'Rainwater harvesting and greywater treatment systems for domestic application in Ireland'. *Desalination*, *260*, 1–8.

McMahon, A. (2015). 'Europe should follow Irish example on austerity, says economist'. *Irish Times, 29 July 2015*. Retrieved from http://irishtimes.com/news/social-affairs/europe-should-follow-irish-example-on-austerity-says-economist-1.2076828 (accessed on 29 June 2016)

OECD. (2012). *Environmental outlook to 2050*. Paris: OECD.

Pape, J., Fahy, F., Davies, A. and Rau, H. (2011). 'Developing policies and instruments for sustainable consumption: Irish experiences and futures'. *Journal of Consumer Policy*, *34*(1), 25–42.

Smith A. and Raven R. (2012). 'What is protective space? Reconsidering niches in transitions to sustainability'. *Research Policy*, 41, 1025–1036.

Verplanken, B. and Orbell, S. (2003). 'Reflections on past behaviour: A self-report index of habit strength'. *Journal of Applied Social Psychology*, *33*(6), 1313-1330.

9

URBAN LIVING LABS, "SMART" INNOVATION AND THE REALITIES OF EVERYDAY ACCESS TO ENERGY

Vanesa Castán Broto

1. Introduction

The development of information and communication technology (ICT), the World Wide Web and the rise of social media constitute some of the key technological developments of our times. These technological developments have also influenced our understanding of cities, urbanisation and its management. ICT emphasises connectivity and so it has led to a greater interest of understanding cities in relation to flows, whether these are material or information flows. The promise of smart cities has inspired the creation of urban laboratories to experiment, specifically, with ideas of "smart governance", as it is the case in, for example in Amsterdam (Meijer and Bolivar, 2016). In this way, smart city ideals have giving new impulse to fantasies of controlling and managing the city.

New understandings of cities have emphasised their increasing complexity and how this is predicated in a complex web of transferences beyond and within distances, or urban teleconnections (Seto et al., 2012). Yet, the recognition of the increased complexity and dynamic character of urban areas, and their characterisation as an open system in continuous interaction with close and distant hinterlands has not daunted planners, civil engineers and consultants to rethink notions of urban management in relation to ICT developments. The innovation is that, unlikely ideas of eco-city, the discourse of smart cities is directly inspired by ideas of competitive growth and economic development and thus, it has captured the imagination of city managers seeking economic growth.

The discourse of "smartness" is just another attempt at regenerating discourses of urban development, despite its presentation as a vision that engages with the full complexity of urban systems (Kourtit and Nijkamp, 2016). Smart cities support visions of urban futures that have inspired new or recycled strategies for ICT-based urban design, planning and management (Angelidou 2015; Firmino and Duarte, 2016). In the context of the global demand for decarbonisation to avoid dangerous climate change, ICT-based strategies for smart cities are most often directed towards the optimisation of energy systems, to improve efficiency and reduce carbon emissions on the one hand, and to facilitate the participation of wider publics through social media and similar tools on the other (Luque-Ayala and Marvin, 2015a).

The question in this paper is the extent to which smart city visions are fit to deliver both environmental sustainability and social justice in contemporary cities. In particular, I am interested in the extent to which plans to make "smart cities" address the everyday concerns of urban dwellers. I approach this question through a retrofitting perspective that engages with the potential to improve service delivery in existing cities (Hodson and Marvin, 2015), leaving aside fantasies about the possibility of creating new smart cities from scratch (Watson, 2014). Urban development projects are never developed on a blank slate: urbanisation depends on the transformation of existing land use relations. While there is a promise embedded in the idea of experimenting with smart cities and creating urban laboratories of smart governance, there are clear challenges for attaining cities that are socially and environmentally just. In many ways, smart city discourses signify a mere regurgitation of ideas of technological urban control and the alignment of planning objectives with those of reproducing market mechanism and economic competitiveness (often associated with surveillance fantasies, as explained in Firmino and Duarte, 2016). Thinking about the creation of operational and citizen-oriented smart cities will depend on the extent to which smart city visions, and the means deployed to deliver them, respond to the needs of cities and their material constraints.

This entails examining those visions against the background of the urban infrastructure landscapes that emerge in cities. Infrastructure landscapes focus on understanding the city as it is experienced by people as emerging from the coevolution of social practices, cultural and technological histories and the transformation of ecologies (Castán Broto, 2017). Urban infrastructure landscapes can be used as an analytical tool to think of what cities actually are, and how visions are matched to existing realities. The analysis follows three case studies in Hong Kong, Bangalore and Maputo. The comparative analysis shows, first, the heterogeneity of perspectives that emerge to characterise smart cities; and, second, their realisation of smartness as an on-going project of urban development which is never really accomplished.

2. Two visions of smart cities

Smart cities is a general concept that is used to ICT-related applications in urban design, planning and management. For many, the notion of smart cities has come to substitute eco-city dreams that were common in the mid-2000s and that came to be implemented in famous examples such as the desert city of Masdar or the now abandoned stalled eco city Dongtan in Choming, Chongming Island, near Shanghai. Eco-cities is a term already used to refer to a variety of approaches to urban management (Caprotti, 2014). Equally diverse is the field of smart cities. The difference is that under a smart city discourse cities are positioned as active agents of economic development (Meijer and Bolivar, 2016). Smart city discourses are often associated with shifts toward entrepreneurial urbanisation (Datta, 2015).

Smart evokes ideas of intelligence and good judgement, alongside ideas of neatness and tidiness. The kind of intelligence that we think of as smart in common parlance is witty and fast: that of an interlocutor who can respond cleverly in real time. These ideas resonate through the smart cities literature in which cities emerge as complex artefacts that can be "taught to think" cleverly, quickly and in real time. Thinking is akin to connecting and accumulating information.

There is an undoubtedly technological focus on the smart cities literature, which evaluates the applicability of such technologies in relation to issues of performance (Caragliu et al. 2011; Lee, Phaal and Lee, 2013; Walravens, 2012). However, smart cities are not simply cities that use advanced ICT technologies. Rather, smart cities are cities that integrate management systems that engage both institutions and citizens in the management of the city: smartness is also about people and about systems of governance (Meijer and Bolivar, 2016). For smart cities' proponents, this is a kind of thinking that relates to cities' ability to advance different aspects of its environment, economy and society, in a kind of "smart everything" discourse. For example, a widely cited early paper on smart cities (Giffinger et al., 2007) identifies six "smartness" areas: (1) "Smart economy" relates to the city competitiveness, the innovative "spirit" of the city, the presence of "smart" (ICT-oriented) industries or the productivity and labour flexibility of the city; "Smart people" related to the social and human capital, for example the extent to which people are educated; "Smart governance" relates to the extent to which people can participate in the governance of the city and transparency; "Smart mobility" relates both to the performance and accessibility of transport systems, including how they facilitate the city"s connectivity; "Smart environment" relates to the use of natural resources in the city, pollution; and "Smart living" relates to the extent to which the city provides for the quality of life of its citizens and the attractiveness of the city.

This rather broad view of smart cities suggests that "smart" pertains every aspect of life in the city, as indeed, ICT and digital technologies have become key tools through which numerous aspects of urban life are managed and dealt with. This broad view of smart cities is common (e.g. Kourtit and Nijkamp, 2016). Yet, in practice, the implementation of smart cities is narrower than it appears in smart city utopias.

The understanding of smart shapes how smart cities are implemented. One way to characterise the "smart" in smart cities is to think about the specific interventions and technologies that come associated to it. Connectivity, clouds, the Internet of Things and other technical terms relate to: (1) the application of ICT to the management of critical infrastructures or (2) the application of ICT to improve the governance of the city (Chourabi et al., 2012). Beyond the context-specific models, there are two fundamental models of ICT applications that emerge closely associated to the two technological models of urban design and management for environmental sustainability, one which we could think of as top-down smart cities, where a centralised governing body organises the network; and another which we could think of as bottom-up smart cities, in which a myriad of citizens or citizens group intervene in a process of distributed control of the network.

Top-down smart cities are those in which the "smart" is related to the application of technologies to monitor large amounts of data (big data) and use such data to optimise resource flows. In relation to an urban energy system, this requires the use of large amounts of data, for example to optimise the flows of energy – especially electricity – at different stages of the electricity network. ICT technologies enable the automation of processes of data collection, with smart metres acting as sensors that compile data of energy consumption and can feed back to the service provider, thus connecting consumer preferences with the provision of services, enable providers to predict peak demands and optimising the whole network. Some use the term "Internet of Things" to emphasise the interconnectivity between different objects in a resource mediating network.

Automated data collection enables urban managers to redesign the city in a more efficient way – the promises are various: reducing carbon emissions, optimising the use of resources, responding more efficiently to consumer needs. The requirements are political leadership, an educated citizenship and enough investment for a network of sensors that can be managed appropriately. These visions are led by companies that provide connectivity services. Barcelona (Spain), for example is working with Cisco's system to implement a "smart city model" that would permeate all its services, including energy and transport. A key element of this plan is the installation of an optic fibre network that enables integration of different networks. Sensor-based information is used, for example, to remotely control street-level lighting, after transitioning 50 streets and more than 1,100 lampposts to LED technology. Working with utilities, the municipality intends to improve energy efficiency through the deployment of more than 19,500 smart metres in the Olympic Villa. Other resource management includes remoted-controlled irrigation of green spaces and fountains and a hot water network to 64 buildings.

This is a technology-oriented, cybernetic vision of urban management. As reported in Cisco's information note, the coordinator of this initiative in Barcelona explains that "This would not have been possible if we did not have top-down political vision . . . You can start thinking bottom-up, but the big, final push was at the political level. If you don't have political willingness, it is impossible".[1] This vision requires the controlled participation of citizens in relation to sensors, to use them appropriately. Educating citizens to support political leadership and perform their lives appropriately, according to the new requirements of integrated networks, is a key requirement for the delivery of top-down visions of smart cities (see, e.g. Belanche, Casalo and Orus, 2016). Citizens participate in the city's governance through remote apps that may improve the transparency of impossibly large, difficult to handle data. Active participation requires a high level of education and capacity to intervene through ICT networks, something that is lacking in most cities. Technological-led perspectives on smart cities have been, however, widely criticised by urbanists engaging with the need for more holistic agendas of urban development (Zubizarreta, Seravalli and Arrizabalaga, 2016). More sophisticated perspectives underemphasise the role of technology to highlight smart city visions in which citizens become active forces for city regeneration through the development of smart institutions (Huston et al., 2015).

In contrast, bottom-up perspectives on smart cities emphasise the use of ICT networks to connect people who, in turn, develop a collective intelligence "programming" the city collectively. In complexity theory, only truly bottom-up technologies are putatively disruptive (Batty, 2016). For example, bottom-up smart city visions relate to the development of open source software online and the development of Internet-based collective action that has reached a wider public through social media. Here, ICT applications are not controlled by expert-led municipal departments or by multi-national private companies, but rather, by groups of ICT oriented programmers and developers who can provide platforms to develop alternative services or keep in check government and business institutions.

This follows a worldwide movement of citizen science, which challenges the nature of expertise and identifies the role of citizens in producing knowledge about their neighbourhoods and environments (Irwin, 1995). Citizen's science emphasises the need for enlarging the processes of knowledge production to include a wider public. In its simplest form, citizen's science involves citizens with simple protocols to collect data observable in their neighbourhood. However, proponents of citizen's science call for recognising the alternative means to interpret reality and formulate questions that emerge from citizen's experiences of their environment (Haklay, 2013). Citizen science concerns is related to deeper questions about the democratic possibilities of a society where there are differentiated attribution of responsibilities and entitlements to produce knowledge. In cities, there is ample potential for distributed forms of intelligence in citizens to challenge the existing ways of operating in a city. By becoming ICT-savvy, citizens challenge the knowledge production process that leads to more, rather than less, participation in decision-making processes.

While this conceptualisation of smart cities relies on similar infrastructure (wi-fi, cloud storage), it is also predicated in a completely different way of understanding internet not as an Internet of Things, connected through sensors and big data, but rather, as independent citizens connected with mobile phones creating shared platforms (Townsend, 2013). Both visions are interconnected and can coexist such it happens in cities like Barcelona. The key difference here is that while the top-down vision of smart cities emphasises the needs for investment in smart networks and citizens' education, the bottom-up vision seeks how to make the most from the growing number of mobile phone users worldwide, even in poorer areas that lack essential infrastructures. The bottom-up version is one in which social media and programming innovations help keeping in check anybody that needs to be

TABLE 9.1 Top-down and bottom-up models of "smartness" for urban planning

Model	Enabling technology	Governance	Objectives/aspirations
Top-down	New ICT technologies and the Internet of Things	Centralised commanding centre	Increasing efficiency and operations of infrastructure
Bottom-up	Increasing access of citizens to social media and networks	Distributed control by citizens and citizens' groups	Deliver services matching local expectations and needs

kept in check whether this is big businesses, the government or communities, while enabling decentralised, post-networked infrastructure to develop. For example, SeeClickFix.com is a local advocacy website in New Haven where users can highlight doorstep issues in Google Maps, hence attracting attention to local problems. Bottom-up smartness may also relate with the creation of incubation labs to let "smart" businesses to emerge or with the possibilities for citizens to agglutinate demand so that they can pool collective infrastructure or keep utilities in check.

What we see in both cases, is the performance of governmentalities of smartness, the search for new means of control, in line with the preoccupation with carbon control and carbon urbanism that has emerged in the last years as a response to climate change challenges (While, Jonas and Gibbs, 2010). Smart city is about creating modes of authority over urban systems that appear uncontrollable. The ultimate aim is to enrol citizens in the production of an efficient city, through processes of subjectification whereby citizens themselves respond to the cues in governmental technologies. The performance of control is linked to the deployment of specific urban governmentalities (Luque-Ayala and Marvin, 2015b). The innovation is that the smart city enrols simultaneously the civil servant, the multinational corporative worker, the creative professional and the hacker through the transformative potential of a group of technologies (although they may differ on the intended impacts of such transformations). The hacker themselves is configured as a subject whose operation is part of processes of algorithmic manipulation at the heart of smart cities.

The two orientations above relate to alternative beliefs about what means of control are more effective with citizens. The top-down perspective presents a view of consumer-citizens whose practices may be inscribed in certain technologies through education and management controls. The bottom-up perspective enrols a wider group of citizens in the governing of the self, so that every participant in the network may be monitoring other citizens. These are not opposed visions, neither are they exclusive. Rather, smartness often comes wrap up in a combination of simultaneous establishment of top-down controls with the generation of a distributed intelligence. Yet, each model has at its core very different assumptions about what is a city, how it works, and what is the relationship between urban society and urban technologies (Table 9. 1). The contrast between these two models of city "smartness" helps evaluate the articulation of smart city discourses in different urban contexts.

3. Urban energy landscapes

Discourses of smart cities have influenced both the generation of ULL as a tool for urban governance and the design of ULL. In line with the objectives of this book, this chapter examines the question of what actually happens when ideas of smartness are put into operation in actual cities. The implementation of smart cities as a discourse for the constitution of ULL has implications for the environmental, social and political sustainability of cities that come to the fore in specific contexts of urban development.

The methodology consists of comparing these two different visions of "smartness" and how they are interpreted in different cities specific urban areas, as they are experienced by people in relation to building or developing low carbon cities. This follows previous engagements with studying "actually existing smart cities", not as *sites extraordinaires* but as visions that are implemented in mundane contexts (cf. Shelton et al., 2015). This also constitutes an incipient attempt to engage with the materiality of smart cities as the urban fabric is actually enrolled, following Bulkeley, McGuirk and Dowling (2016). Urban energy landscapes refer to the matrix that enables certain social practices that rely on energy, in a particular spatial context and with dependence of specific energy flows. Urban energy landscapes integrate spatial experiences, cultural histories and material transformations. Looking at urban energy systems in this way moves away from the system of provision and demand and looks instead into the historical, spatial and ecological factors that explain a particular configuration of energy uses and spatial practices. In this vein, urban energy landscapes consist a collection of socio-energetic relations which can be observed on narratives of energy and artefact histories.

There are three theoretical pillars to understand UELs: (1) a phenomenological perspective on landscape that build on Heidegger's notion of dwelling as "being-in-the-world", away from an utilitarian sense of landscape, but in a landscape that emerge from routine engagements with the surroundings and their manipulation with artefacts in a myriad of purposeless but orchestrated interactions, in which the multiplicity of purpose confounds the overall purpose of the landscape (following Ingold, 2001); (2) a cultural-historical perspective on landscape that emphasises their evolution in relation to changing relations with particular artefacts and their design, and the transformation of those visions through the integration of knowledge and ideas in local to global flows and (3) an urban political ecology perspective that emphasises the mutual dependence of the power structures and the transformation of ecologies and characterise resource transformations in relation to particular urban infrastructure landscapes (Gandy, 2011).

The objective of this paper is to understand the extent to which smart city visions are fit to deliver both environmental sustainability and social justice in contemporary cities, particularly in relation to urban energy systems. The results start with a brief characterisation of urban energy landscapes that involve reflecting upon their cultural history, how they are interpreted, their experiences and the material transformations that relate to those histories and spatial experiences. This includes analysis of several sources including qualitative interviewing of representatives of key organisations in the city, participatory mapping and discussions with citizens, with a special focus on difficulties to access energy services, and walking transects to characterise the spatial conditions in which energy is provided and used.

To understand the impact of the implementation of smart cities ideas in such energy landscapes, I ask three questions:

1 How are smart city narratives integrated within existing governance frameworks and discourses in that particular urban energy landscape?
2 To what extent would smart city improvements deliver energy services as they matter to people?
3 Will smart cities work in the actual material contexts in which they are implemented?

The results are presented as follows: first, each urban energy landscape in Hong Kong, Bangalore and Maputo is characterised through a snapshot that highlights its key aspects; then, each of the three questions above is answered in turn. The final section provides a comparative assessment of the different views of smartness and how they relate to the contexts in which they emerge.

4. Comparative analysis of "smartness" in Hong Kong, Bangalore and Maputo

4.1. Urban energy landscapes in three cities

The former British colony of Hong Kong has become a prosperous city propelled first by international trade and later, but the consolidation of an international financial centre in one of the most liberalised markets of the world. The return of Hong Kong to the People's Republic of China in 1997 did not challenge this status as Hong Kong was awarded special economic status, but over the years, mainland Chinese have invested in Hong Kong property and shape its economy in routine encounters from the provision of energy and water resources to frequent shopping trips by Chinese tourists. Attempts to align the political system in Hong Kong with the one in Beijing have led to political tensions as exemplified by the civil disobedience of the Umbrella Revolution in 2014, which despite its worldwide coverage did not bring about significant political changes. Disappearing technologies that embody the tradition of the former city, such as neon, have been taken as symbols of the underlying tension between the liberal and democratic Hong Kong and the communist, innovation-oriented models imposed by Beijing.

Environmentally, Hong Kong has a peculiar status because of its achievements in terms of sustainability and urban planning. Hong Kong is "low carbon by chance", because its low carbon per capita ratio relates to the long-term urban development of the city over several decades, following efforts to resettle people from informal settlements in social housing in the New Territories, away from the city's centre, which led to the development of an effective and dense public transport network and the consolidation of a compact city model of development, with mixed use areas. Central areas like Mong Kok are among the densest populated areas in the world. Simultaneously, the energy system of Hong Kong relies on fossil fuels, especially coal and gas imports from Indonesia and Australia, which supply electricity and gas for the whole island. With extremely low tariffs in relation to average salaries, Hong Kong relies on a 10-year Scheme of Control Agreements that benefits the two utilities in the city, but which constitutes a disincentive for efficiency and energy saving measures.

Energy practices are largely shaped by the architecture of the city, and the pressure over space. Space has the highest premium in Hong Kong, and while growing high has resolved problems of space, it has also resulted in complex demands for energy, particularly for air conditioning. The individual air conditioning unit dominates the landscape, as a reflection of the enduring cultural practices of cooling spaces. Hong Kong is a cool place, where air conditioning has a large share of energy consumption. The architecture supports individual air conditioning units, and they pile high in streets and houses. Energy access is not dependent on the coverage of the network, which is almost universal, or the affordability of energy prices, because of the relative cheap tariffs. Rather, energy poverty results from lack of space and poor appliances. For example, for those living in the poorest houses, such as in illegal converted warehouses with almost no space, coordinating different energy-dependent activities such as cooking, studying or watching TV depends on carefully planned choreographies and forms of coordination between all the members of one family.

Bangalore is also a former British colony, but post-colonial development has happened in an entirely different context. The post-independence period was characterised by the development of household-based industries. Since the 1980s, however, Bangalore has become a world centre for the IT and offshoring industries. This has led to the development of a new class of cosmopolitan Bangaloreans, which may come from outside the city, and who have different demands in terms of the development of the city, its environmental conservation and its consumer habits. Since the appointment of Prime Minister Narendra Modi in 2014, the discourse of smart cities has been elevated, albeit incoherently, to national urban policy (Hoelscher, 2016).

Unlike Hong Kong, Bangalore has a very complex energy supply system that involves numerous sources of power, including hydropower plants, state-owned thermal power plants, nationally owned fossil-fuel and nuclear plants and some marginal renewable resources. The coordination of demand, transmission and power production requires the coordination of different bodies. BESCOM is the key actor involved in securing the city's electricity supply. Bangalore currently suffers from power shortages. The Times of India reports frequent complaints from citizens because of power shortages that BESCOM blames on needed maintenance work, but intermittency also relates to the overall shortfall of energy in the State, which from time to time affects the city. The relative unreliability of the system in comparison to citizens' expectations, especially for those who form part of the cosmopolitan class, has led to numerous private solutions to deal with energy shortages and unreliable services.

On the one hand, for those who can afford it, securitisation takes place with the use of advanced technologies and diesel generators. This is particularly prominent among accommodated classes who are leading a process of peri urban urbanisation in which design and technology is dedicated to the securitisation of resources. For the lower classes, low tech solutions have enabled off-grid and cheap solutions. This is manifest, for example in the fast spread of solar water heaters among the middle classes as a means to guarantee the provision of hot water. Among informal settlement dwellers, social enterprises such as SELCO India have promoted individual solar technologies and improve cookstoves as a means to ensure energy access in a landscape in which informal settlement dwellers are most likely forgotten.

The energy demand of the city is shaped by the increasing demand of a growing industry of IT and offshoring, which has erected itself as a beacon of sustainability. On the streets, a more pragmatic landscape of energy use dominates, with the opportunistic use of the centralised network and individual private solutions. One key aspect of energy use in Bangalore is the ubiquitous presence of unsafe practices of distribution in which individuals connect to the grid through non-regulated connections, while simultaneously relying on private diesel generators.

Maputo represents a completely different energy landscape, in that electricity and gas are not the main sources of domestic energy, but rather, most people rely on charcoal as the main fuel for cooking and use electricity mostly for lighting and communication purposes. Mozambique has abundant energy resources with one of the greatest hydropower plants (Cahora Bassa) in Africa and abundant fossil fuel reserves. However, the electrification rate is very low and even in its capital, Maputo, the central division of Kampfumo has full coverage of electricity. Access to electricity has improved with the implementation of a pre-paid system through which local people can control how much electricity they use and fraction the payments in relation to resources available (Baptista, 2016).

Yet, most families continue cooking with charcoal. The whole architecture of informal settlements in Maputo is geared towards the use of charcoal, and multiple local economies depend on this fuel (Castán Broto et al., 2015, 2014). Here, cookstoves constitute the key artefact that shapes the energy landscape, shaping the flows from the distant forests where charcoal is produced, the central markets where it is distributed and the street corners where mostly women sell small fractions of charcoal in a quantity that reflects local needs. Recent development efforts have tended to connect households to the electricity network and improve cookstoves performance.

These cases represent three different snapshots of urban energy landscapes. Each urban energy landscape is of course dependent on the history and place where it emerges. However, by looking at this contrasting cases we can see a typology of urban energy landscapes emerging in relation to the patterns of provision (type of fuel, distribution system) and the patterns of use and energy access (spatial inequalities in energy access). Smart city ideals play out very differently in each of these contexts.

4.2. Smart city visions and urban realities

We start our assessment in relation to the extent to which smart cities visions play out in government's narratives of the city. Both Hong Kong and Bangalore are thought of as examples of smart city, but the city they portray is a very different one. Led by the Hong Kong SAR Government, albeit through a limited company, smart city strategies in Hong Kong are directed towards developing new economic markets and reindustrialising the city. The smart wearables industry is one that has developed in relation to Hong Kong's Cyberport. The Cyberport is:

> a creative digital cluster with over 600 community members. It is managed by Hong Kong Cyberport Management Company Limited which is wholly owned by the Hong Kong SAR Government. With a vision to establish itself as a leading information and communications technology (ICT) hub in the Asia-Pacific region, Cyberport is committed to facilitating the local economy by nurturing ICT industry start-ups and entrepreneurs, driving collaboration to pool resources and create business opportunities, and accelerating ICT adoption through strategic initiatives and partnerships. Equipped with an array of state-of-the-art ICT facilities and a cutting-edge broadband network, the Cyberport community is home to four grade-A intelligent office buildings, a five-star design hotel, and a retail entertainment complex.

While there is an interest in developing "smart city" solutions, other means to achieve sustainability through energy policies remain limited to traditional approaches. There is now a city-wide consultation to see what will be the follow-up of the Scheme of Control Agreements and whether there is an opportunity to diversify the energy sector. However, diversification is not directed towards sustainable fuels as it is widely assumed that there is no space in Hong Kong for renewables; equally, efficiency campaigns such as those to limit the use of air conditioning, have had little impact so far. The discourse of smartness is one centred on economic competitiveness and promoting new "smart industries". This may have consequences for the energy system in terms of energy demand but it does not seem directed towards improving the use of resources or using more sustainable fuels.

The smart city vision of Bangalore is very different. Local authorities (ULBs) have little influence in comparison with the national government and state institutions and private initiatives. In the case of Bangalore, smart cities discourses emerge in diverse forums from the national government to local private actors. The national government has promoted a smart-city strategy that equates private sector participation with citizens' participation and emphasises the use ICT for sustainable innovation. A key aspect is the attempt to reinforce local governance through citizens' participation through ICT technologies. However, citizens' participation is narrowly conceived and sometimes it has been thought of as token consultation.

On the other hand, there is a strong push from the private sector to implement and celebrate smart city initiatives. These do not emerge as city-wide attempts to create smartness – such as in the case of Hong Kong, even if such smartness is limited – but rather to confined attempts to think of the city as smart. Cisco Smart City, for example, is a blueprint for a campus, which attempts to improve the environmental and social performance of gated communities associated with the IT and offshoring industries. Here, energy securitisation, with the control of energy use in smart buildings and the optimisation of energy networks, is central for the delivery of this blueprint. The networked campus is a marketing tool for the implementation of a particular urban vision that emphasises the possibility to facilitate the communication between a control centre and distributed actors. This is offered as a model of communication between government and citizens, but one which emphasises remoteness and distance.

Finally, in Maputo, the discourses of energy governance emphasise the immediate problems of energy supply in the city and how they are progressively met by different actors. While the public utility Electricidade de Moçambique (EDM) is keen to highlight the achievements of the prepaid system, local NGOs and community leaders emphasise progress in delivering immediate services. The concerns of those trying to improve the situation of cooking with charcoal, from reducing indoor pollution to preventing children's accidents, remains far from the smart city rhetoric. Instead, the significant advances that mobile phones have brought to people, as a means of employment and social relations, and their active participation in social networks are the best signs of bottom-up smartness that defies the homogenising conceptualisations of smartness that dominate the literature of smart cities.

The three cases question to what extent smart city visions correspond to the needs of citizens in each city. In the case of Hong Kong, economic development imperatives are in stark contrast with the current limitations of the energy system, the constraints to switch to a more sustainable supply system with less dependence on imports from fossil fuels and the extent to which the precarious situation of poorer people and the way energy poverty is produced within a socio-technical system as understood within government authorities. The problem of housing is also one of energy as housing determines the ever-increasing energy needs, while the actual conditions of housing explains why certain cultures of energy use are reproduced. In the case of Bangalore, the question is whether technology-led projects represent a large enough part of the population, and the extent to which this is providing a solution for those in small industries, those who rely on illegal connexions in energy networks and those who live in informal settlements and still depend on woodfuel or coal for cooking. While in Bangalore there is a clear policy incentive to push for smart cities, even when this happens in campuses and separated from the rest of the city, the case of Maputo begs the question of what a smart city rhetoric can offer for cities where the majority of the population has precarious access to services. The possibilities of organising communities through social media both for protesting the lack of services and for coordinating community-based services hold certain promise but will require accessible platforms – and crucially energy – to develop fully. This appears to be a chicken and egg kind of problem, where smartness cannot emerge without better conditions for energy access.

Finally, the three cases illustrate the material limitations to deliver smart cities. In the case of Hong Kong, the compact city structure and land prices drive poverty and access to good services, not because electricity is not cheap enough but because there is no space to use appropriate appliances or to enjoy the services associated with energy. Cheap electricity alone does not guarantee quality of life because electricity is an enabler for social practices, and the use of electricity is not a means in itself. Why does it matter that I can turn on the TV when I am sharing the room with my son who needs silence and light to study? In contrast, in Bangalore, the indomitable city is tamed by fencing out certain areas of privilege where smartness is controlled, as much as its citizens. Fragmentation, rather than connection, is what fosters innovation in Bangalore's landscape. Finally, in places like Maputo, where services are lacking, smartness means very little, other than the possibilities opened for collective action in the face of governance failures. People succeed routinely in engaging with material possibilities to deliver sustainable livelihoods, despite the lack of services or the lack of government capacity to deliver well-being improvements.

4.3. Top-down versus bottom-up perspectives on smart cities

These are idiosyncratic examples that depend on the context and history of the city. Nevertheless, certain patterns begin to emerge in relation to how discourses are enacted and performed. Table 9.2 summarises the comparative analysis by highlighting the top-down and bottom-up components in each case and the extent to which the point toward different typologies of "smartness".

The case of Hong Kong is a top-down model but strongly oriented not towards improving the management of resources, but towards the development of a "smart" economy. The bottom-up element emerges in the focus on entrepreneurship as a means to develop innovations, but the constraints for this innovative spirit to emerge are clearly set by enabling policies such as the Cyberport, which guide and censor smartness possibilities. Smart resides in the character of the industry, rather than any other characteristic.

TABLE 9.2 Comparative analysis of the articulation of discourses of smartness in each case study

Smart city vision	Hong Kong	Bangalore	Maputo
Top-down	Top-down, control-based ideas at odds to the city's ideology of freedom, as political difference from PRC	Top-down visions are limited to gated compounds and campuses where their materiality can be (relatively) controlled	Any top-down attempts at implemented ICT appear as fantasies in the current I/F context
Bottom-up	Government-enabled incubators for start-ups promote bottom-up but within a hegemonic view of smart economies	Spontaneous development of "smart", socially-based innovation for the poor	Potential for social media based solutions to provide alternatives for energy access and activism

In Bangalore, there is an orientation towards networked technologies and optimisation of energy systems, but, to enrol the material in an effective way, smartness is confined spatially, in campuses where technologies can be demonstrated. The bottom-up component emerges in relation to the lack of communication between government and citizens, and the possibility to mediate that remotely.

Finally, in Maputo, top-down versions of smartness are limited, and in any case, directed towards building resilience and improving the provision of basic services, for example, through the prepaid system. There is little emphasis on the quality of life of citizens, other than coping with the provision of basic services. Yet, in Maputo there is the greatest potential for truly bottom-up forms of smartness, because communities already organise themselves and mobile phones are ubiquitous. Finding the means to connect this to service provision is not easy, but a certain possibility in the absence of governance.

Read together, the three cases point towards smart city as an idea that means different things simultaneously. There is potential for the reinvention of smartness in different contexts, and the appropriation of technology for progressive aims, particularly where it is not immediately obvious how to implement smart alternatives. However, the proliferation of discourses of smartness as opposed to eco-cities or sustainable cities represent an attempt to put a particular kind of technology at the centre of urban planning, thus engaging with what is, in essence, technocratic forms of governance. This alignment with the deployment of smartness governmentalities as a means to control and enrol populations in existing regimes of hegemony raises suspicions about whether any kind of progressive ideal can be advanced through smart visions.

5. Conclusion

Back to the original question, are we likely to envisage an urban revolution associated to smart city discourses in ULL? ULL have been advocated as creative spaces where multiple actors can work together to deliver sustainable urban innovation. This book shows how some of those attempts are linked to efforts to bring in citizens' participation in urban planning and deliver impacts on the ground that affects people's well-being and urban ecosystem. Are smart city visions fit to inform the development of ULL? Can these smart city prescriptions for urban planning help deliver environmental sustainability and social justice? This would mean delivering smart city that simultaneously improve the use of resources and meet citizens' needs. Meeting citizens' needs, for example should mean ensuring energy access for basic services for the poor.

There are indeed vibrant debates of what smart cities are or should be. While some claim to encompass almost any aspect of sustainable cities, the truth is that most ideals of smart city can be summarised in two presumptions- on the one hand there is a top down model for the promotion of "smart technologies" or the connection of networks with sensors to optimise the use of resources; on the other there is a bottom-up model that emphasises civic activism through social media, for the provision or monitoring of services. However, alternative storylines of smartness are necessary to counter the corporate lobby that has so far been allowed to shape this discourse (Söderström et al., 2014). The connection of "smartness" with alternative economies and ideas of sharing seems to be a plausible avenue for such alternatives (McLaren and Agyeman, 2015). Participatory understandings of the smart city are also necessary to unlock the sustainable promise of "smartness", if there is one (Aylett, 2015).

These are different examples, selected for their representation of a variety of energy landscapes, but they show how the notion of smart cities is likely to play very differently in different contexts. In most cases, it is likely to emerge in consonance with existing processes of urban development, whether this is by emphasising smart as supporting private businesses in Hong Kong, as connecting with the standard trend towards campus living in Bangalore or as being absent in the case of Maputo.

In each case, these are narrow conceptions of what is to be smart. Smart is ultimately about being connected and communicating appropriately. Whether is thought of in top-down or bottom-up ways, the underpinning assumptions that are most often deployed as smart urbanism are closely similar (Luque-Ayala and Marvin, 2015a). There is a vague promise of progressive urban policy in bottom-up discourses of smartness, but in practice, such promises have never been realised. Obsessions of efficiency and control largely dominate the delivery of smart cities in a coordinated way. Perhaps the operation of ULL relates less to the implementation of well-defined vision and more to the analysis of what conditions enable multiple forms of innovation. Yet, in increasingly unequal cities, there is a risk that any innovation-led approach will exacerbate inequality. As explained in the example above, networked campuses and smart economic areas increase inequalities and limit spatial patterns of resource access privilege. One way to rethink the notion of smart cities would be to examine which technologies and artefacts make it more connected and communicated. My interest is on cities that are not planned as smart, but rather, where "smart" is an emergent process and where connectivity promotes innovation, but hinders control. The question that cities should answer is how they can use the increase connectivity between citizens to return authority and powers, alongside with management responsibilities, to citizens themselves.

Note

1   Cisco's Barcelona jurisdiction profile is available on their website: www.cisco.com/ assets/global/BE/tomorrow-starts-here/pdf/barcelona_jurisdiction_profile_be.pdf (last accessed 20 February 2018).

Acknowledgements

This chapter is an output of a project called Mapping Urban Energy Landscapes, funded by the Economic and Social Research Council (Grant reference: ES/ K001361/1). This chapter is dedicated to the memory of Alex Aylett.

References

Angelidou, M. (2015). 'Smart cities: A conjuncture of four forces'. *Cities*, *47*, 95–106.
Aylett, A. (2015). Green cities and smart cities: The potential and pitfalls of digitally-enabled green urbanism. *UGEC Viewpoints*. Retrieved from https://ugecviewpoints.wordpress. com/2015/06/11/smart-green-cities-can-we-enable-deeply-sustainable-urbanism-through-new-media-technologies/.
Batty, M. (2016). 'How disruptive is the smart cities movement?' *Environment and Planning B-Planning and Design*, *43*(3), 441–443.
Belanche, D., Casalo, L. V. and Orus, C. (2016). 'City attachment and use of urban services: Benefits for smart cities'. *Cities*, *50*, 75–81.
Bulkeley, H., McGuirk, P. M. and Dowling, R. (2016). 'Making a smart city for the smart grid? The urban material politics of actualising smart electricity networks'. *Environment and Planning A*, *48*(9), 1709–1726.
Caprotti, F. (2014). 'Critical research on eco-cities? A walk through the Sino-Singapore Tianjin Eco-City, China'. *Cities*, *36*, 10–17.
Caragliu, A., Del Bo, C. and Nijkamp, P. (2011). 'Smart cities in Europe'. *Journal of Urban Technology*, *18*(2), 65–82.
Castán Broto, V. (2017). 'Energy landscapes and urban trajectories towards sustainability'. *Energy Policy*, *108*: 755–764.
Chourabi, H. et al. (2012). 'Understanding smart cities: An integrative framework'. ed. *System Science (HICSS), 2012 45th Hawaii International Conference on*, 2012, 2289–2297.
Datta, A. (2015). 'New urban utopias of postcolonial India: "Entrepreneurial urbanization" in Dholera smart city, Gujarat'. *Dialogues in Human Geography*, *5*(1), 3–22.
Firmino, R. and Duarte, F. (2016). 'Private video monitoring of public spaces: The construction of new invisible territories'. *Urban Studies*, *53*(4), 741–754.

Giffinger, R. et al. (2007). *Smart cities – Ranking of European medium-sized cities*. Vienna: Vienna University of Technology.

Haklay, M. (2013). 'Citizen science and volunteered geographic information: Overview and typology of participation'. In D. Sui, S. Elwood and M. Goodchild (Eds.), *Crowdsourcing Geographic Knowledge* (pp. 105–122). Dordrecht, The Netherlands: Springer.

Hodson, M. and Marvin, S. (2015). *Retrofitting cities: Priorities, governance and experimentation*. London: Routledge.

Hoelscher, K. (2016). 'The evolution of the smart cities agenda in India'. *International Area Studies Review, 19*(1), 28–44.

Huston, S., Rahimzad, R. and Parsa, A. (2015). '"Smart" sustainable urban regeneration: Institutions, quality and financial innovation'. *Cities, 48*, 66–75.

Irwin, A. (1995). *Citizen science: A study of people, expertise and sustainable development*. New York: Psychology Press.

Kourtit, K. and Nijkamp, P. (2016). 'Exploring the "New Urban World"'. *Annals of Regional Science, 56*(3), 591–596.

Lee, J. H., Phaal, R. and Lee, S.-H. (2013). 'An integrated service-device-technology roadmap for smart city development'. *Technological Forecasting and Social Change, 80*(2), 286–306.

Luque-Ayala, A. and Marvin, S. (2015a). 'Developing a critical understanding of smart urbanism?' *Urban Studies, 52*(12), 2105–2116.

Luque-Ayala, A. and Marvin, S. (2015b). 'The maintenance of urban circulation: An operational logic of infrastructural control'. *Environment and Planning D: Society and Space, 34*(2), 191–208.

McLaren, D. and Agyeman, J. (2015). *Sharing cities: A case for truly smart and sustainable cities*. Cambridge, MA: MIT Press.

Meijer, A. and Bolivar, M. P. R. (2016). 'Governing the smart city: A review of the literature on smart urban governance'. *International Review of Administrative Sciences, 82*(2), 392–408.

Seto, K. C., et al. (2012). 'Urban land teleconnections and sustainability'. *Proceedings of the National Academy of Sciences, 109*(20), 7687–7692.

Shelton, T., Zook, M. and Wiig, A. (2015). 'The "actually existing smart city"'. *Cambridge Journal of Regions Economy and Society, 8*(1), 13–25.

Söderström, O., Paasche, T. and Klauser, F. (2014). 'Smart cities as corporate storytelling'. *City, 18*(3), 307–320.

Townsend, A. M. (2013). *Smart cities: Big data, civic hackers, and the quest for a new utopia*. New York: WW Norton and Company.

Walravens, N. (2012). 'Mobile business and the smart city: Developing a business model framework to include public design parameters for mobile city services'. *Journal of Theoretical and Applied Electronic Commerce Research, 7*(3), 121–135.

Watson, V. (2014). 'African urban fantasies: Dreams or nightmares?' *Environment and Urbanization, 26*(1), 215–231.

While, A., Jonas, A. E. and Gibbs, D. (2010). 'From sustainable development to carbon control: Eco-state restructuring and the politics of urban and regional development'. *Transactions of the Institute of British Geographers, 35*(1), 76–93.

Zubizarreta, I., Seravalli, A. and Arrizabalaga, S. (2016). 'Smart city concept: What it is and what it should be'. *Journal of Urban Planning and Development, 142*(1), 8.

PART III
Processes of ULL

10

15 YEARS AND STILL LIVING

The Basel Pilot Region laboratory and
Switzerland's pursuit of a 2,000-Watt
Society

*Gregory Trencher, Achim Geissler and
Yasuhiro Yamanaka*

1. Introduction

The effectiveness of global and national responses to climate change largely
hinges on the collective contributions of individual cities (Bulkeley, 2015; Trencher
et al., 2016). Accordingly, cities are increasingly expected to pave the way by
experimenting with and demonstrating innovative technologies and social
arrangements addressing sustainability challenges. Urban living laboratories (ULL)
are a powerful enabling paradigm to this end (Voytenko, McCormick, Evans and
Schliwa, 2016). Historically, scientific experimentation was limited to the confines
of laboratories or field observations (Evans and Karvonen, 2011). In ULL however,
messy, uncontrollable and "living" urban environments are targeted by ambitious
cross-organisational collaborations to trial and deploy cutting-edge technical and
social solutions to advance societal progress towards greater sustainability and
prosperity (Evans, Karvonen and Raven, 2016; Voytenko et al., 2016).

Scholarship portrays important characteristics defining ULL. They are physically
embedded in geographical locations (Voytenko et al., 2016), usually at the city or
neighbourhood scale. Often, they carry a focus on the built-environment (Evans,
2016; Evans and Karvonen, 2014), energy, ICT (Mulder, 2012; Veeckman and
van der Graaf, 2015) or mobility (Joller and Urmas, 2016). Reflecting wider
scientific shifts towards mode two science[1] and transdisciplinarity (Karvonen and
van Heur, 2014), they typically involve collaboration across multiple societal
sectors such as academia, government, industry and citizenry (Juujärvi and Pesso,
2013). As such, ULL perform multiple services for multiple stakeholders. They
provide: scientific evidence from observation and application of knowledge (Evans
and Karvonen, 2014); understanding into how new technologies and services
interact with human systems and the conditions required for spurring wider

diffusion (Joller and Urmas, 2016); training, education and new business development opportunities for industry and citizens (McCormick and Kiss, 2015); policy test beds for municipalities, and importantly, integration of users (e.g. citizens) into the innovation process (Voytenko et al., 2016). Often, they are laden with high political expectations for reviving stagnant urban spaces and driving economic development (Evans, Bulkeley, Voytenko, McCormick and Curtis, 2016). The future orientation of ULL also provides policy makers and scholars the precious opportunity to reframe urban development narratives towards prosperity, sustainability and technological progress. Experimentation (i.e. trials of novel technologies and social arrangements) is central to ULL (Heiskanen, Jalas, Rinkinen and Tainio, 2015; Karvonen and van Heur, 2014). This experimentation ethos is propelled by assumptions that old patterns are unsustainable, that new is smarter, and, since arguably nobody yet knows how an authentically sustainable urban environment might look and function, much trial and error is needed to get there. Although ULL are generally greeted with positivity by scholars and practitioners alike, some raise concerns about social inclusiveness, equal distribution of benefits, and generally unquestioned acceptance of neoliberal development paradigms, albeit with a green lining (Curtis, 2015; Karvonen, Evans and van Heur, 2014).

Scale is also integral to the ULL's premise. Focusing attention on a single city, neighbourhood, building or sector (e.g. food or energy etc.) encourages risk-taking and ambition since resources can be concentrated, leading to higher visibility of projects and easier monitoring of outcomes than if working on a larger scale. Arguably, the ultimate ambition of the ULL is to spur imitation by other players, thus triggering potentially important societal shifts on a wider scale (Sengers, Berkhout, Wieczorek and Raven, 2016).

With ULL appearing mostly over the last decade (Dutilleul, Birrer and Mensink, 2010), knowledge is limited on long-term societal impacts and their capacity to trigger broader societal change. Understanding also lacks on potential life-cycles by which they may form, mature and evolve. Filling this gap, this chapter examines 15 years of experiences accumulated in the 2,000-Watt Society Basel Pilot Region (2,000WS-BPR) in Switzerland. Since 2001, this lab has united scientists with government and industry practitioners to exploit the Canton Basel-Stadt as a testing arena for emerging built environment, mobility and energy technologies to advance progress towards a "2,000-Watt Society" (elaborated below).

To further understanding on ULL design approaches and processes, we examine three key time phases in 2,000WS-BPR's development to understand:

1   Design strategies of the laboratory, and factors affecting its societal legitimacy
2   Processes for fostering experimentation and sustainability innovation in industry, resulting impacts, and factors positively or negatively affecting the lab's capacity to advance progress towards a 2,000-Watt Society
3   Learning and adaptive measures taken after a decade of experimentation and implementation.

Our data were collected over four years (2012–2016). Primary data was obtained from fifteen semi-structured interviews (via telephone or in person in Zurich, Basel and Boston) with participating university, government and industry stakeholders. Secondary data were obtained through academic publications, press articles and project documentation.

2. 2,000-Watt Society and analytical framework

While many societies struggle to articulate scientifically and socially legitimate visions of sustainability and take concrete steps towards their materialisation, Swiss society has enthusiastically adopted the target of 2,000 watts of continuous power per citizen as the guiding principle for long-term energy and sustainability strategies. Alongside Basel, this target has been adopted by the Swiss Federal Office of Energy, the cities of Zurich, Geneva and the majority of individual towns and villages across Switzerland (Stulz, Tanner and Sigg, 2011). In Zurich's case, a referendum has resulted in 76.4 per cent in favour of a legally binding obligation to attain the 2,000-watt target by 2050 (City of Zurich, 2011).

The scientific vision of a 2,000-Watt Society emerged from the Swiss Federal Institutes of Technology (ETH) domain[2] in the early 1990s. Its forefathers (Imboden, Jaeger and Müller-Herold, 1992; Kesselring and Winter, 1994) argued that industrialised nations should take steps to reduce over several decades their mean permanent flux of non-renewable primary energy (consumed in buildings, transport and manufacturing of materials and goods) to around 2,000 watts per citizen. This corresponds to an energy demand of roughly around 17,500 kWh or 1.5 tonnes of oil equivalent per capita per year. Conceived as a dual response to climate change and mounting security concerns over limited fossil fuel supplies, 2,000 watts per citizen mirrored the world average in the nineties and, incidentally, the Swiss level of energy demand in 1960.

The vision of a 2,000-watt future focuses on two key indicators: energy demand and GHG emissions (Stulz et al., 2011). This concept therefore demands both energy efficiency and a transition to renewable energy. On one hand, attaining this target in Switzerland would necessitate slashing energy demand – equivalent to around 6,000 watts of continuous power demand in the year 2000 (Morosini, 2010) – by two-thirds, and over a timescale of fifty to 150 years (Jochem, 2004). On the other hand, sourcing seventy 5 per cent of a 2,000-watt energy demand from renewables would cut GHG emissions to one tonne of CO_2e per capita per year. This was seen as largely compatible with an emissions trajectory for limiting the post-industrial global temperature rise to within 2°C.

While the still evolving qualitative concept of sustainability was shrouded in ambiguity, the quantitative notion of a 2,000-Watt Society won support from the presidential board (ETH-Rat) of the universities and research institutes comprising the ETH domain. An ETH faculty respondent recounted that this was due to its simplicity and objectiveness as a measure for sustainability. In 1998 ETH-Rat

INTENSITY OF ETH/NOVATLANTIS ROLE IN
IMPLEMENTATION

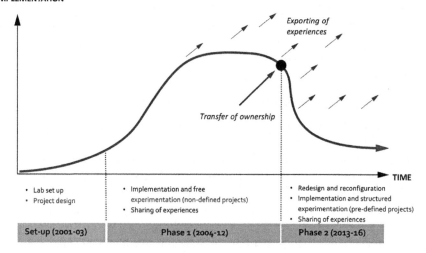

FIGURE 10.1 Lifecycle of the 2,000WS-BPR

formally endorsed pursuit of this target through public statements, fixing the 2,000-Watt Society as the focus of the ETH domain's sustainability strategy (Marechal, Favrat and Jochem, 2005). The scientific vision of a 2,000-Watt Society thus shifted from a purely scientific vision to an agenda for implementation, which later prompted establishment of the 2,000WS-BPR.

The analytical framework for examining the lifecycle of the 2,000WS-BPR is shown in Figure 10.1. This depicts approximately 15 years of experiences accumulated over three key phases: 1) design and set-up, 2) initial implementation, and 3) reflection, restructuring and re-implementation. Time is represented on the x-axis, with the y-axis reflecting the degree to which ETH and Novatlantis (an ETH sustainability platform) directed the lab and played a role in implementation projects. Each phase is examined individually and our temporal examination of each reveals distinct qualities about the ever-evolving design, structure, processes and impacts of the lab. Moreover, their ensemble generates important insights into how ULL can evolve and adapt over time. Emphasis is given to the transfer of "ownership" of the lab to the Canton Basel-Stadt at the end of the first phase. This is to elucidate a key point we later emphasise. That is, in addition to triggering environmental and societal transformations, a crucial output of ULL can be the continuation of the governance structure itself, as the university-initiated lab is "spun-off" to society after initial university leadership and funding lessens. An important legacy of ULL can thus be a governance framework for uniting and guiding long-term collaborative efforts to advance societal sustainability across scientific institutions, government and industry.

3. Set-up phase (2001–2003)

The structure of the 2,000WS-BPR essentially consists of a long-term partnership connecting scientists from the ETH domain and the University of Applied Science and Arts Northwestern Switzerland (FHNW) with planners and administrators from the Canton of Basel-Stadt. Project design and co-ordination in the lab occurs mostly through this collaboration involving scientists and municipality practitioners. Stakeholders from local industry (e.g. vehicle fleet managers, construction companies, architects, professional associations etc.) and energy utilities are then recruited to assume implementation. 2,000WS-BPR differs somewhat from ULL conceptions that emphasise citizen engagement (Bergvall-Kareborn and Stahlbrost, 2009; Franz, 2015) since they are not explicitly incorporated into project design. Yet citizens play a passive – albeit important – role by using or engaging with emerging mobility and building technologies and providing feedback to researchers.

For the first decade, this science-society interaction and overall direction of 2,000WS-BPR was facilitated by Novatlantis. This platform was conceived in the year 2000 by the ETH domain as a mechanism for carrying out transdisciplinarity, which was attracting much interest in ETH at the time (Klein et al., 2001). Novatlantis's initial mandate was simply to take "results of recent research within the ETH domain" and apply them to "projects designed to promote sustainable development in major urban settlements" (Novatlantis, 2013) in tandem with societal stakeholders. In 2001 Novatlantis fixed its attention on the 2,000-Watt Society as its flagship "product". Efforts were made to translate this from a scientific vision into an implementation and societal transformation project.

The City of Basel (situated in Canton Basel-Stadt) was selected as the pilot site for a scaled-down application of the 2,000-Watt Society vision, which at the time, was a simultaneously global and national scale concept. Economically prosperous Basel is currently home to some 186,000 residents and the third largest city in Switzerland, after Zurich and Geneva. The objective of the lab was to spur national progress to a 2,000-Watt Society by trialling and demonstrating in social settings a suite of emerging technologies from the ETH domain, and by integrating scientific knowledge into industry construction projects. Projects in the lab are concentrated within the jurisdictional boundaries of Canton Basel-Stadt (37 km^2). The Canton's participation was conditional. It stipulated strongly that projects and experiments implemented in the lab carry a strong practical focus.

2,000WS-BPR adopted a three-pronged focus on individual buildings, large-scale urban development and individual mobility. Within each foci, multiple pilot projects (i.e. permanent and pioneering built environment projects by industry) and demonstrations (i.e. temporary scientific mobility experiments) were conceived through a bottom-up approach. Project conception occurred through workshops uniting canton planners and industry ("the demand" side) with scientists and researchers (the "supply side").

The set-up phase shed much light on factors that can challenge the legitimacy of ULL projects designed by scientific and municipality actors, and potential strategies to overcome these. On the ETH side, despite general enthusiasm for the 2,000-Watt Society vision, Lienin, Kasemir and Stulz (2004) and Novatlantis actors state that some researchers were highly concerned about the seemingly "unscientific" nature of implementation-orientated projects, and potential problems they might pose for the basic science culture and reputation of ETH. To address this, two strategies were adopted. First, needs for scientific robustness were addressed through a quality control mechanism. ETH members on the 2,000WS-BPR steering committee screened implementation project proposals for approval and allocated seed funding, in accord with scientific merit. Second, Novatlantis researchers and project managers were obliged to contribute to publications in peer-reviewed outlets. Despite these measures, several respondents reflected that priorities in basic research and scientifically orientated projects with publication potential significantly hampered efforts by Novatlantis to recruit ETH scientists into implementation-orientated projects.

When it came time to scientists and practitioners to co-design projects, contrasting orientations and motivations significantly limited the ability of initial workshops to bear fruitful ideas for implementation projects with scientific value. Figure 10.2 depicts the nature of the two paradigm polarities encountered, as described by a FHNW researcher. Academic researchers had priorities in conducting research on scientific agendas with an extremely long-term focus, and as mentioned,

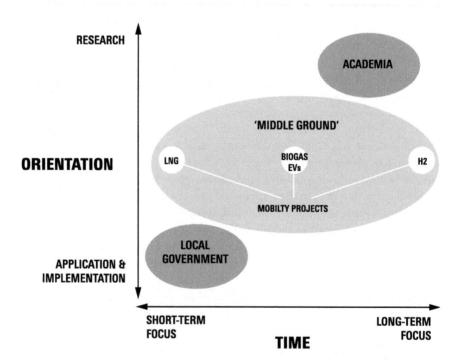

FIGURE 10.2 Paradigm polarities between academia and local government. After Binz (2013)

potential publication value. In contrast, city planners lived in a world of implementation where resource restraints and four-year budget and political cycles foster a short-term focus. These opposing cultures stifled the formation of a common agenda, to the extent that, as a FHNW researcher recalled, "it was like we were from two different planets". After much persistence and several workshops, a strategy emerged of co-designing projects to fall into the "middle ground". That is, projects that served differing time horizons, and generated both scientific and implementation value. In mobility, Lienin, Kasemir, Stulz and Wokaun (2005) describe how pilot and demonstration projects were designed to address time horizons that were *short-term* and practical (LNG engines employing commercially available technologies and fuelling infrastructure), *mid-term* and emerging (EV's and biogas engines) or *long-term* and visionary (hydrogen fuel cell vehicle prototypes). Practical value for industry and government stakeholders was enhanced by trialling vehicle technologies in existing, large-scale industry and government car fleets. Conversely, scientific value was assured by simultaneously monitoring technical performance and socio-economic aspects such as costs and public acceptance, and by integrating future orientated research agendas such as hydrogen. Similarly, for built environment projects, a programme for fostering individual pilot buildings addressed implementation needs in the short-term. On the other hand, large-scale urban development projects addressed the mid-to long-term, since large projects require relatively more time for planning and construction.

4. Phase one (2004–2012): Initial implementation

Discussion will now turn to noteworthy processes employed at both the local and national level in the first implementation phase to carry out experimentation and foster sustainability innovation in industry, resulting impacts, and driving or hampering factors.

To realise a fleet of cutting-edge pilot buildings demonstrating various innovations in energy efficient construction and engineering, a "P+D" (pilot and demonstration) programme was established. This ensemble of pilot buildings aims to showcase previously unattained standards in sustainable building innovation and inspire replication across industry. This strategy consisted of a) FHNW faculty identifying candidate buildings (in new construction and renovation) from permit applications submitted to the Canton that demonstrated commitment to energy efficiency, b) contacting and encouraging owners and architects to increase energy efficiency through integration of additional construction techniques and technologies, and c) incentivising this through financial subsidies from the Canton to offset a portion of additional costs, and permitting use of the brand "2,000-Watt Society pilot building". To spur ambition and innovation, allocation of subsidies sometimes demanded commitment to attaining standards such as Minergie-P[3] or zero-energy in addition to minimal use of construction materials, and future-orientated designs. In other cases, subsidies focused on bringing architects and engineers to trial emerging building materials and technologies such as vacuum insulation panels (vacuum sealed and high-performance insulation), LUCIDO facades (for heating buildings with solar energy) and co-generation systems (i.e. combined heat and power).

This approach fostered a portfolio of approximately thirty individual demonstration buildings across the city (Canton of Basel-Stadt and Novatlantis, 2013). Each demonstrates new levels of achievement in aspects such as building envelope design, construction materials, insulation, heating and cooling systems, and onsite renewable energy. This portfolio serves as an important learning mechanism for the building community regarding technical and economic dimensions of low-carbon building innovation. It is expected, a driver for replication. Learning is facilitated through regular onsite tours for engineers, architects and developers in demonstration buildings of interest. Building users and engineers are also able to provide designers first hand feedback on comfort, performance and maintenance. Nationally, experiences and good construction practices are shared through annual forums for the building industry. For the public, P+D programme buildings form a tangible beacon of progress towards the energy efficiency requirements of a 2,000-Watt Society.

As other processes, governance structures controlling or influencing construction industry behaviour were shaped to spur innovation in low-carbon building at both the local and national level. In the language of transitions theory (Loorbach, 2014), this corresponds to a novel approach of tinkering with the "regime" (i.e. national rule structures) to spur innovation at the "niche" level (i.e. localised, individual projects).

Locally, a key approach involved incorporating 2,000-Watt Society principles into rule structures governing large-scale urban development projects. Canton-owned land in several locations was pitched to the private sector for development through competitive bidding. Novatlantis and FHNW actors participated in the judging panel. This enabled the awarding of tenders to development proposals best confirming to strict sustainability and energy efficiency principles (see Fischer, 2009) based on a 2,000-Watt Society. Consultations with Novatlantis practitioners and FNWH scientists throughout planning helped assure that these were subsequently implemented into awarded projects.

Over the past decade, four large-scale urban development projects were fostered across Basel-Stadt. In the spirit of the pilot building programme, these aim to set a new standard for future urban development trajectories by demonstrating previously unattained levels of innovation, energy efficiency, and environmental sustainability. The most recent is Erlenmatt West; a 19 hectare mixed redevelopment of a former Deutsche Railway freight yard, where construction began in 2013. The development comprises commercial and residential buildings, aged care and shopping facilities, restaurants, schools and parklands. Buildings boast features such as Minergie[4] certification, advanced insulation and energy efficient heating, minimalist facades and use of recycled concrete to reduce embedded energy in construction materials, and smaller than average living areas to reduce energy consumption and increase rental affordability. Compatibility with the

2,000-Watt Society is communicated prominently in marketing, and most notably through procurement of a "2,000-Watt Society neighbourhood-level" certification (managed by the Swiss Federal Office of Energy). Erlenmatt West thus contributes significantly to mainstreaming and increasing the visibility of the 2,000-Watt Society concept in the public and political realm. Equally, it has set a new industry standard for low-carbon innovation and integration of holistic sustainability planning into large-scale urban development, both locally in Basel and nationally. It is expected that the neighbourhood 2,000-Watt Society certification will drive widespread imitation of the new standard across industry.

As a national strategy, 2,000-Watt Society related sustainability and energy efficiency knowledge (Jochem, 2004) together with first-hand building experiences from the pilot region were integrated into engineering guidelines by the Swiss Society of Engineers and Architects (SIA, 2011). The "Energy Efficiency Roadmap SIA 2040" outlines a multi-decadal pathway for the construction industry to raise energy efficiency in buildings. This was possible, since the managing director of Novatlantis at the time, together with ETH faculty, served on the SIA commission. By redefining industry norms and desirable benchmarks, this strategy guides low-carbon and holistic sustainability planning in construction projects both across Switzerland, and locally in Basel. As an intermediate goal, this energy efficiency roadmap for industry fixes the year 2050 to attain 2,000-Watts of non-renewable primary energy use and two tonnes of CO_2 per capita per year in (i) *construction* (embodied energy in building materials) and daily energy consumption, (ii) *operation* (energy consumption in heating, cooling, hot water, lighting and electricity) and (iii) *mobility* (transport patterns of building users in residential, offices and school sectors). In parallel, an Excel calculation tool was developed for industry to facilitate estimations of energy efficiency off construction plans.

Integration of 2,000-Watt Society energy efficiency principles into industry governance frameworks at the local and national scale has contributed largely to standardising, mainstreaming and institutionalising sustainability and energy efficiency concepts in Swiss new construction. For example, several individual buildings in Zurich, Basel and Geneva have used the Excel tool and energy efficiency standards outlined in the 2040 path. Additionally, Swiss Cantons and cities across Switzerland also base local building energy codes upon SIA guidelines. In parallel, the City of Zurich and Swiss Federal Office of Energy have also published precise guidelines for calculating 2,000-Watt Society projects (2,000 Watt Bilanzierungkonzept) across Switzerland. This approach has also driven low-carbon innovation in 2,000 WS-BPR itself. For example the aforementioned West Erlenmatt development integrates 2,000-Watt Society standards and principles specified by the SIA to guide construction planning. This nurtured a holistic pursuit of sustainability and energy efficiency in construction and urban develop-ment that encompasses construction materials and waste, onsite electricity production and usage, and transport.

A third process for fostering sustainability experimentation involved long-term public demonstrations of low-carbon automobile technologies in existing vehicle fleets in industry and the municipality. By integrating mobility technologies developed in ETH research institutes, these experiments involved ETH scientists and an explicit research dimension. With ambitions to encourage replication by other industry fleet managers and politicians, roughly half a dozen projects were implemented over a decade (Canton of Basel-Stadt and Novatlantis, 2013). These demonstrated the technical and economic feasibility of various sustainable transport solutions for the short term (natural gas), mid-term (biogas and EV) and long term (hydrogen fuel cells).

A landmark mobility experiment involved a six-month trial in 2009, then one-month in 2011, of a hydrogen fuel cell vehicle (HFCV) street sweeper developed in ETH research institutes (EMPA and PSI). Called hy.muve, this prototype was integrated into the Canton's diesel-operated street sweeper fleet. Integration of a HFCV prototype into the existing fleet served a promising way to literally showcase on the sidewalks hydrogen safety while generating insights into future options for addressing the fleet's diesel-reliance and air-pollution and GHG emission contributions. This experiment provided important learning into both technical and social dimensions of hydrogen mobility technologies, which at the time, were largely unproven. Knowledge gained on technical dimensions included system performance, handling, wear and tear (inclusive of a real life breakdown necessitating the hauling of hy.muve back to ETH for repairs), energy efficiency and GHG emissions.

Importantly, technological demonstration projects generated rich learning into social aspects. For the hy.muve, learning encompassed public acceptance, affordability and regulatory requirements for permitting public operation of HFCVs and fuelling stations. Importantly, the hy.muve experiment also allowed learning into how the various technical and human elements comprising a hydrogen transportation system (HFCVs, drivers, hydrogen manufacturers, fuelling stations etc.) would interact. As emphasised by an ETH scientist; "you just can't learn about this sort of stuff in the conventional laboratory". Learning from the hy.muve field testing was also used to shape new research agendas for ETH researchers. In recent years, they have shifted their focus away from research and development and technological performance testing towards analysis of socio-economic opportunities and barriers in diffusing HFCVs, and demonstration of integrated vehicle and fuelling infrastructure systems.

A notable strength of this lab stems from incorporation of the scientific 2,000-Watt Society concept as its core guiding principle – or *leitmotif*. This serves as a "boundary object" (Clark et al., 2016) allowing stakeholders from differing societal sectors to share a common vision, target and language. It also serves as an intellectual glue binding the individual building and mobility projects together into an integrated portfolio. Also important, the 2,000-watt continuous power and corresponding one tonne CO_2 per capita per year target mutually function as an objective and quantifiable yardstick for sustainability. In contrast, the fuzzy and entirely qualitative concept of "sustainability" offers no such objectiveness – since its core weakness is the lack of a widely accepted quantitative definition (Kharrazi,

Rovenskaya, Fath, Yarime and Kraines, 2013). Furthermore, the quantitative nature of the 2,000-watt vision assures that progress towards this target is *measurable*. This facilitates the fixing of mid-term targets and roadmaps to monitor progress. The aforementioned SIA Energy Efficiency Roadmap, for attaining 3,500 watts (comprising 2,000 watts from non-renewables) and two tonnes of CO_2 per capita (from a 50 per cent reduction in fossil fuels) per year by 2050, is one such example. In addition, the multi-decade to multi-century vision timescale of the transition to a 2,000-Watt Society leads naturally to long-term thinking and partnerships. This is a key driver behind the continuation of 2,000 WS-BPR even after 15 years.

As other strengths, the explicit orientation towards implementation rather than research (highlighted during the design phase) is highly significant. Although this dampened scientific interest in the lab in some quarters of the ETH domain, the emphasis on practice was central in attaining societal legitimacy and securing the necessary participation and funding from local government and industry to bring projects to fruition (Eberwein, Lobsiger-Kägi, Eschenauer, Jetel and Carabias, 2015). Integration of scientific energy efficiency principles into implementation projects was also a natural outcome of allowing industry practitioners in both Novatlantis and the consulting firm to play a key role in shaping and driving projects. Without practitioners at the helm of the pilot region, projects might have run off in a completely research-oriented direction. This would undermine industry absorption of scientific knowledge and subsequent societal impact.

Despite such strengths, several shortcomings in the first decade generated important learning opportunities for 2,000WS-BPR actors. The first and probably most significant concerns the limitations of the paradoxical approach of attempting a social transformation with predominantly technical experts and technical means. As lamented by Morosini (2010, p. 66), "most communication on the 2,000-Watt Society focuses on watts, not on society. Technological change is to the fore, not social change". One illustration of the pitfall of single-mindedly pursuing energy efficiency through technological improvements without regard to lifestyles is the "rebound effect". In residential buildings, this occurs when increased average flooring areas or preferences for higher indoor temperature settings offset efficiency gains in lighting, heating or building design. This has indeed occurred in Switzerland. From 1955 to 2005, average per capita living space has doubled (City of Zurich, 2008). This predicament has prompted the emergence of the value-laden concept of "sufficiency" in Zurich and national discussions on the 2,000-Watt Society (City of Zurich, 2011; Notter, Meyer and Althaus, 2013). Sufficiency reflects growing scientific and political acceptance that attaining a 2,000-Watt Society will eventually require individual citizens to curb tendencies towards greater energy consumption through larger living and working spaces and higher material consumption. Although BPR-2,000WS actors are still unsure of how to promote sufficiency through technical innovation, the above-mentioned West Erlenmatt development (through smaller than average dwelling sizes) and SIA guidelines (calling for stabilisation of average per capita flooring) suggest that urban environments can at least be engineered to foster sufficiency. Secondly, learning emerged at the end of the first decade that not enough had been done to integrate research into the portfolio of demonstration buildings and urban development. The pendulum, seemingly, had swung too far in the direction of implementation without regard for monitoring the actual energy performance of completed construction projects. The educational value of the laboratory had suffered, since it did not collect data to establish which forms of construction innovation lead to high energy efficiency gains. Thirdly, success was acknowledged in advancing sustainable construction practices locally and nationally from shaping building industry governance structures and implementing multiple pilot buildings and urban development projects. However, a strong sentiment emerged that the real challenge now lay in how to drive retrofitting in existing buildings.

5. Phase two (2013–2016): Redesign and reconfiguration

Coinciding with the retirement of key protagonists and exhaustion of ETH-funding support, 2,000WS-BPR underwent a major restructuring and redesign at the end of the first decade. This lead to a second and re-imagined four-year phase of implementation for 2013–2016.

Similar to the first decade, a focus was retained on buildings and mobility. However, a third theme of renewable energy was incorporated. This was largely in response to heightening global concerns about energy security, particularly following the Fukushima nuclear disaster in 2011, and a renewed push towards renewables in the EU. This resulted in a triple pilot region focus on *construction*, *energy* and *mobility* (both individual and public). Another noteworthy shift is a change of focus from new construction to retrofitting. This addresses above-mentioned learning that more efforts were required to raise the energy efficiency of the existing building stock, inclusive of many historical buildings.

In terms of basic governance structure, 2,000WS-BPR remains a three-party science-government collaboration between Novatlantis (representing ETH research institutes), FHNW and the Canton Basel-Stadt. As opposed to phase one, project design occurred prior to the beginning of the phase at the level of scientific research institutions and the municipality, while implementation is again carried out by industry. As another fundamental shift, ownership of the lab was "spun off" from the ETH domain to the Canton of Basel-Stadt's Office of Environment and Energy (see Figure 10.1). As such, the Canton now assumes overall funding, promotion and co-ordination of the pilot region. Accordingly, the role of Novatlantis and ETH in the overall direction of the lab and implementation activities has decreased, relative to earlier years. No longer funding or directly co-ordinating the pilot region, the major role of Novatlantis and ETH now concerns project management and guidance in overall steering, together with FHNW and the Canton.

In terms of lab design, the most fundamental shift to occur was a switch to a portfolio of project agendas predefined by the steering committee. These were conceived mostly by scientific actors from ETH/Novatlantis and FNWH in response to, on one hand, priority areas identified in collaboration with Canton planners, and on the other hand, current research agendas and promising mobility and building technologies developed in the ETH domain. This led to a preliminary portfolio of around twenty projects, of which a dozen were chosen and approved by the Canton. This approach contrasts to the first decade, where project ideas were sourced from the bottom-up, through workshop driven brainstorming sessions between ETH scientists and city planners and administrators. As seen in Table 10.1 and Box 10.1, the lab has now adopted an approach of "outsourcing" to industry the implementation of pilot projects focused on the trial, demonstration and diffusion of specific emerging technologies with sustainability advancing potential. Each project has a specific description, objective, expectations and a designated scientific project manager from FHNW, ETH or the consulting firm. A seed-fund budget also accompanies each project. These serve as incentives and subsidies to offset some of the additional costs or financial risks for industry ensuing the implementation of unproven technologies.

This change in strategy mirrored various perceived advantages and learning accumulated from the first decade. First, pre-defined projects allowed the steering committee to fix visions, approaches and technologies that were pioneering, ambitious and not yet in industry's or the public's field of vision. Experiences in the first decade showed that bottom-up sourced projects tended to lack ambition since implementing industrial partners tended to focus on less risky, commercially proven technologies. Second, this approach allowed adjustment at the macro steering level to ensure greater cohesion and complementarity between individual projects. It also permitted a more integrated and holistic approach where energy and buildings, or energy and mobility, were tackled in tandem rather than in isolation, as in the first decade. This approach is particularly salient in the mobility projects. Here a shift has occurred from evaluating technical feasibility of individual vehicle technologies through temporary demonstration projects to assembling commercially viable business models integrating hydrogen vehicle fleets, fuel production and fuelling station infrastructure. Finally, pre-defined research agendas allowed the scientific institutions to integrate an explicit research element into projects, particularly with construction and energy. This was aided by establishing a budget to support an array of specific research projects (e.g. Energy Hub in Table 10.1). Such strategies proved a significant step in overcoming the competing tensions between scientific needs for research value, and industry and municipality needs for implementation value (see Figure 10.2). This also addresses the previous lack of measures to document the learning and energy performance gains of individual pilot buildings, and share these with the building community.

This pre-defined project approach, however, is not a problem free approach to designing ULL. Since implementation hinges upon successful recruitment of the right industrial partner, project implementation is, in many cases, proving slow. Considerable efforts are required for scientific project managers to recruit and then "educate" industrial partners on the social and economic significance of emerging

TABLE 10.1 Overview of key projects in phase two

Project name/ area	Project type (and implementation status)	Objective
Construction		
Spray-on aerogel insulation	Demonstration (partly achieved)	Showcase spray-on, insulating aerogel plaster (developed at EMPA) to raise energy efficiency of old, heritage-listed buildings
Prefabricated facade modules	Demonstration (not achieved)	Showcase prefabricated, high-insulation building facades with integrated mechanical ventilation ducts for energy efficiency renovations on walls and roofs with minimal site disruption
Energy		
Coloured module PVs	Demonstration/ research (achieved)	Showcase/measure technical performance of coloured PV modules on building exterior walls
Second-life batteries	Demonstration/ research (achieved)	Showcase/measure technical and economic potential of used EV lithium ion batteries to store excess onsite renewable energy in buildings
Energy hub	Research (achieved)	Investigate impacts/benefits of integrated, decentralized and adaptive renewable energy systems for cities
Solar power to methane	Demonstration (not achieved)	Explore potential of converting excess solar power to synthetic methane gas and hydrogen
River water use	Research/feasibility study (achieved)	Examine theoretical/technological/ economic potential of using river water from the Rhine for district heating and cooling in residential estates
Mobility		
Electromobility	Demonstration (a) achieved, (b) under planning	(a) Showcase/examine potential of electric buses to reduce GHG emissions and fossil fuel use in public buses and (b) install fast-charging public stations in Basel
Hydrogen fuelling station	Demonstration (under planning)	(a) Retrofit gasoline station to provide hydrogen to commercial fleet of fuel-cell vehicles, and (b) showcase potential of mixing renewably sourced hydrogen in LNG for vehicles
Gas-battery hybrid vehicles	Demonstration (not achieved)	Showcase/examine potential of gas (LNG or biogas) and battery hybrid vehicles (developed at EMPA) for commercial vehicle fleets

---

BOX 10.1 COLOURED SOLAR PVs AND SECOND LIFE BATTERY ENERGY STORAGE PILOT PROJECT (PHASE TWO)

As a symbolic gesture of progress towards a 2,000-Watt Society, in spring 2015 a former coal silo at Gundeldinger Feld in Basel was retrofitted to an office building that pilots two emerging building technologies: (1) coloured, building-integrated solar PV panels, and (2) a battery storage system.

Building heritage protection regulations were navigated to install 172 panels on walls and the rooftop (Müller and Steinke, 2015). The coloured PV panels used technology originated from ETH Lausanne. These resemble wall claddings more than conventional counterparts, opening the door to wider applications than "ugly" black panels. To ascertain varying degrees of performance (affected by colour), data are collected at 15-minute intervals, then analysed. Although electricity generation performance is around 5–10 per cent inferior to regular panels, this is offset by a greater surface area than rooftop installations alone.

In parallel, a "second life" battery energy storage system was installed on site, utilising both new and old EV batteries. Also extending the useful life of used EV batteries, the system absorbs excess local solar electricity production, especially on weekends and in summer. Data are collected and analysed to monitor battery behaviour and compare old battery performance to new.

---

technologies and negotiate needs for additional private investments amidst uncertainty around paybacks. Consequently, several projects in Table 10.1 are yet to see the light of day (e.g. prefabricated facade modules and solar power to methane), or are still under planning with industry (e.g. hydrogen fuelling station). Also, despite the resolve to systematically monitor, document and diffuse learning outcomes across the construction industry, seemingly, some building owners do not share aspirations for using their tenant-occupied assets as scientific testing sites. Such an episode occurred in an aerogel insulation renovation project (see Table 10.1). Consequently, the actual energy performance of this completed project today remains unclear. Nevertheless, photographers were commissioned by the Heritage Protection Department of Basel-Stadt to capture various stages of the construction process and demonstrate to industry and policy makers the potential of this technology to enhance the energy efficiency of heritage protected buildings.

Interestingly, one strategy was formulated in the second phase to address the predominantly technical focus of the first decade and the need to tackle the lifestyle sector. The City of Basel appropriated the EU Energy Neighbourhoods model as an attempt to increase public engagement around the 2,000-watt target. Residents compete on a neighbourhood basis to reduce home energy consumption by 9 per cent relative to the previous year, over four months. Although not an integral thematic agenda of 2,000WS-BPR, the Canton of Basel-Stadt has nevertheless made an explicit link in this initiative to the 2,000-watt target and a required public shift towards greater residential energy efficiency. It also demonstrates a first and useful strategy for engaging the lifestyle sector in pursuit of a 2,000-Watt Society.

Despite these adaptive measures, an undeniably technocentric focus still dominates. As shown in Figure 10.3, the 2,000-Watt Society agenda has success-fully evolved from a scientific concept and knowledge base to a knowledge implementation and societal transformation project. However, despite awareness in key 2,000WS-BPR actors of the limitations of predominantly technical approaches (Eberwein et al., 2015; Stulz et al., 2011), the lab is yet to move towards a model of scientific experimentation and industry implementation projects that incorporate both technical and social dimensions. The approach still resembles somewhat of a "let's build hardware first and worry about software later" mentality. As such, the second phase of projects remain largely focused on fostering technological verification and pilot projects of which the primary recipients of learning outcomes are industry and government technicians, rather than citizens. This is not to discount their educational value. Nor is this to dismiss their importance in laying the seeds for greater systematic change of built environment and mobility systems that attainment of the 2,000-watt goal would ultimately require. This continuing technical focus can be easily explained by several factors. These include the engineering and basic science culture of ETH from where the 2,000-Watt vision originated; the technical disposition of the key players and institutions driving the pilot region; and the continued absence of civil society

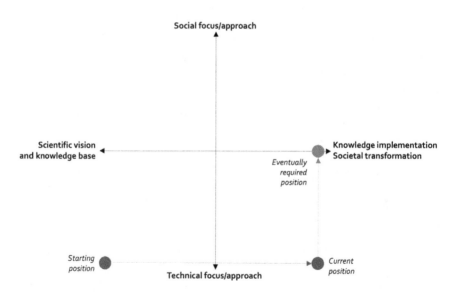

FIGURE 10.3 Trajectory of the 2,000WS-BPR

organisations in the steering committee. Perhaps just as importantly, the technical approach also testifies to the relative ease of implementing scientifically and politically legitimate forms of innovation over value-laden initiatives seeking to tackle the way people live.

6. Conclusion

This chapter examined 15 years of experiences accumulated in the 2,000-Watt Society Basel Pilot Region, Switzerland. This case generated rich insights into the potential long-term impacts of ULL, and outcomes of differing approaches to spurring experimentation and innovation. We set out to build understanding on potential ULL lifecycle patterns, processes for fostering sustainability innovation and societal transformations, long-term impacts, and strengths and shortfalls of distinct approaches and lab designs. Several important observations can be made.

In terms of lifecycles, this lab is an example *par excellence* of a scientific vision and knowledge base evolving into an implementation and societal transformation project. After a decade of projects, the university-originated lab was spun off, transferring official ownership, funding and public endorsement responsibilities to the Canton of Basel-Stadt. This suggests that if tangible results are achieved and benefits made visible to city governments, a key output of ULL can be the creation of spin-off governance platforms that unite scientific institutions, city municipalities and industry in the long-term pursuit of innovation and progress to sustainable urban development. This has clear implications for designing other ULL. *Is this spin-off approach replicable, especially in municipality settings less endowed than Basel?* After all, this was Switzerland. Not only does the City of Basel have cash, they are extremely serious about integrating science into long-term energy and sustainability policies. They also inscribed their commitment to the 2,000-Watt Society in successive legislative plans since 2009. For ULL (and indeed other innovation strategies for industry) seeking longevity and long-term political support to achieve multi-decade transition goals and the formal institutionalisation of municipal support into policy documents appears essential for weathering potentially unpredictable and changing priorities of four-year political cycles.

During the lab design phase, contrasting values and motivations of the university and city planners (termed "paradigm polarities") posed significant challenges. These were overcome by designing projects to fall into middle ground, with both practical and scientific value as well as relevance to differing time horizons (short-term, long-term etc.). However, efforts to spur sustainability innovation in building in the first decade saw the pendulum swing too far towards implementation. Monitoring and evaluation – the *forte* of scientists – of building energy performance was neglected. Consequently, the pilot and demonstration building program lacked a research dimension to verify actual energy performance and document learning. Its educational value for architects and engineers was thus compromised. Addressing this deficiency by integrating research into construction and energy demonstration projects in the second phase was a noteworthy adaptive response. This required a renewed appreciation for scientific knowledge production and a shift from the City of Basel's initial first decade demand for implementation-orientated projects with "no research, no theory" (Eberwein et al., 2015).

2,000WS-BPR demonstrates a sophisticated array of potential processes for ULL to foster experimentation and technical innovation. For mobility, these comprised of demonstration projects integrating emerging low-carbon vehicle technologies in real-world fleets of taxis, buses, municipal street sweepers, and company vehicles. For ETH researchers, such experiments advance scientific understanding by generating valuable knowledge on both technical performance and social dimensions such as the type of fuelling infrastructures, legislation and business models needed for a transition to hydrogen or other low-carbon fuels. As emphasised, such learning is not possible in a four-walled laboratory. In the built environment, municipality subsidies were exploited to foster a portfolio of thirty plus individual buildings showcasing innovation in energy efficiency and sustainable construction. Standing as tangible beacons of progress towards a 2,000-Watt Society, these set new industry benchmarks for sustainability innovation. Onsite tours for industry professionals ensure promotion of learning and imitation across the building community. Finally, frameworks from the 2,000-Watt Society were integrated into national-level governance structures for the building industry. At the local level, these were exploited to guide the holistic pursuit of sustainability and energy efficiency in large-scale urban development projects. At the national level, these enabled the mainstreaming of standards and ambitious new norms for sustainable, energy efficient construction in line with the vision of a 2,000-Watt Society. This approach of tinkering with the regime to influence both project-specific and national building practices has high relevance for transitions theory. Recently, scholars are taking interest in top-down strategies for fostering innovation (Loorbach, 2014), in addition to traditional framings of bottom-up, niche-level innovation. With new construction in Switzerland now largely controlled through these industry norms and building codes, renovation of existing buildings set a new challenge for current pilot region projects. In the second phase, the approach of using government subsidies to foster industry uptake of promising, but unproven, technologies from the ETH domain continues. However, in mobility projects, a notable shift has occurred away from isolated trials of individual technologies. The new approach consists of fostering new holistic business-models and socio-technical configurations comprising of hydrogen and hybrid vehicles, fuelling infrastructure and energy production. This suggests a useful funnel approach of trialling and demonstrating a wide array of promising technologies in early years, and then narrowing in to commercialise and diffuse a selected few in later years.

Evans (2016) highlights limitations inherent to "the episodic nature of one-off projects" in ULL. Clearly, without a common purpose and set of values, a lab such as 2,000WS-BPR risks breaking down into a geographically and temporally scattered portfolio of research and pilot projects that are forgotten once funding runs out and initial momentum wavers. Yet undeniably, a firm glue binds together the myriad of mechanisms and projects executed over 15 years in Basel. It is the vision of a 2,000-Watt Society. Its strength lies in its value as a boundary object, providing a common language and goal for scientists, practitioners and city planners. It serves a dual role as an objective measure of sustainability, and a guiding narrative or *leitmotif*. "Stories" embedded into energy strategies are central determinants of societal buy-in (Janda and Topouzi, 2016). Yet generally, storytelling is not the *forte* of scientists. However, the vision of a 2,000-Watt Society is one of the best sustainability stories out there. Rather than short-term disruptive change, it calls for long-term, incremental improvement. It fits in perfectly with the modernist idea of progress. It is admirably compatible with capitalist society's needs for constant growth, technological advancement, and looking to the future rather than the past.

Yet no sustainability story is perfect. At its core, the 2,000-watt story is essentially about watts and energy efficiency. Influenced by this, 2,000WS-BPR chose a technocentric approach of building hardware first, leaving software for later – a common tendency for the ULL paradigm (Franz, 2015). A solid head start has been made thanks to technology. Yet continued progress towards a 2,000-Watt Society largely depends on the lab's capacity to broaden its pursuit of technical innovation to encompass lifestyle and social dimensions. As contended by Notter et al. (2013), only if people are "educated and governed in order to develop a more sustainable lifestyle based on sufficiency" might a 2,000-Watt Society be eventually possible. As we emphasised however, citizens are not a central actor in this ULL, especially from a project design perspective. Although we offered several explanations for this technocentric focus, active engagement of citizens and lifestyle dimensions into project design and implementation appears an outstanding future challenge. This said, the lab's somewhat top-down approach to engineering urban environments to facilitate lifestyles with higher energy efficiency is certainly an important process for driving sustainability transitions.

Still alive after 15 years, the 2,000WS-BPR deserves careful, continued monitoring. Not overlooking the endowed societal, political and financial resources fuelling progress towards a 2,000-Watt Society in Basel and Switzerland, their ability to achieve this goal will be a solid indicator of the effectiveness of ULL for sparking long-term societal transitions to a low-carbon, sustainable future.

Acknowledgements

We extend thanks to the editors and James Evans for providing valuable feedback. Thank you also to the many pilot region actors cooperating for interviews and providing information. We extend special appreciation to Roland Stulz for continued sharing of experiences over the years and helpful feedback on the manuscript.

Notes

1 This term depicts applied and multidisciplinary scientific knowledge production, implemented with stakeholder involvement, and in response to societal needs.
2 This refers to the two universities (Zurich and Lausanne) and four research institutes comprising ETH.
3 This industry building standard developed in Switzerland corresponds roughly to a "passive house" level in Germany. Its basic principle lies in minimising heat loss through advanced thermal insulation, airtight building shells and ventilation system with heat recovery. Buildings attaining this level have an annual fuel consumption in oil equivalent of approximately three litres per square metre for space heating.
4 This building standard is based on the same principles as Minergie®-P, but with a minimum energy efficiency performance of approximately 4 litres of oil equivalent per square metre of space heating.

References

Bergvall-Kareborn, B. and Stahlbrost, A. (2009). 'Living Lab: An open and citizen-centric approach for innovation'. *International Journal of Innovation and Regional Development, 1*(4), 356–370.

Binz, A. (2013, February). *Interview by author, 11 Feburary, Interviewer: G. Trencher.* Zurich, Switzerland.

Bulkeley, H. (2015). 'Can cities realise their climate potential? Reflections on COP21 Paris and beyond'. *Local Environment, 20*(11), 1405–1409.

Canton of Basel-Stadt and Novatlantis. (2013). *2000 Watt society: 10 Years in the Basel pilot region* (in German). Retrieved from www.novatlantis.ch/wp-content/uploads/2014/11/10-jahrepilotregionweb_wie_02-01.pdf (accessed 24 December 2016).

City of Zurich. (2011). *On the way to the 2000-watt society. Zurich's path to sustainable energy use.* Zurich: Office for Environmental and Health Protection Zurich.

Clark, W. C., Tomich, T. P., van Noordwijk, M., Guston, D., Catacutan, D., Dickson, N. M. and McNie, E. (2016). 'Boundary work for sustainable development: Natural resource management at the Consultative Group on International Agricultural Research'. *Proceedings of the National Academy of Sciences, 113*(17), 4615–4622.

Curtis, S. (2015). 'Innovation and the triple bottom line: Investigating funding mechanisms and social equity issues of living labs for sustainability'. Master of Science in Environmental Management and Policy, Lund University, Lund, Sweden.

Dutilleul, B., Birrer, F. A. J. and Mensink, W. (2010). 'Unpacking European living labs: Analysing innovation's social dimensions'. *Central European Journal of Public Policy, 4*(1), 26.

Eberwein, G., Lobsiger-Kägi, E., Eschenauer, U., Jetel, M. T. and Carabias, V. (2015). Subtask 2 – Switzerland: The 2000 Watt Society. Retrieved from www.ieadsm.org/wp/files/Subtask-2-Switzerland-2000-Watt-Society.pdf (accessed 24 December 2016).

Evans, J. (2016). 'Placing low carbon transitions: Learning to retrofit in laboratories'. In M. Hodson and S. Marvin (Eds.), *Retrofitting cities: Priorities, governance and experimentation* (pp. 195–211). Oxford, UK, and New York: Earthscan.

Evans, J., Bulkeley, H., Voytenko, Y., McCormick, K. and Curtis, S. (2016). 'Circulating experiments: Urban living laboratories and the politics of sustainability'. In A. Jonas, B. Miller, K. Ward and D. Wilson (Eds.), *Handbook on spaces of urban politics.* London: Routledge.

Evans, J. and Karvonen, A. (2011). 'Living laboratories for sustainability: Exploring the epistemology of urban transition'. In H. Bulkeley, V. Castan Broto, M. Hudson and S. Marvin (Eds.), *Cities and low carbon transitions.* New York: Routledge.

Evans, J. and Karvonen, A. (2014). ' "Give me a laboratory and I will lower your carbon footprint!" Urban laboratories and the governance of low-carbon futures'. *International Journal of Urban and Regional Research*, *38*(2), 413–430.

Evans, J., Karvonen, A. and Raven, R. (Eds.). (2016). *The experimental city*. Oxford, UK, and New York: Routledge.

Fischer, S. (2009). '2000-Watt society: the Swiss vision for the creation of sustainable low energy communities'. Paper presented at the 45th ISOCARP Congress 2009, Porto, Portugal.

Franz, Y. (2015). 'Designing social living labs in urban research'. *Info*, *17*(4), 53–66.

Heiskanen, E., Jalas, M., Rinkinen, J. and Tainio, P. (2015). 'The local community as a "low-carbon lab": Promises and perils'. *Environmental Innovation and Societal Transitions*, *14*, 149–164.

Imboden, D., Jaeger, C. C. and Müller-Herold, U. (1992). 'Projekt Energieschranke'. *GAIA – Ecological Perspectives for Science and Society*, *1*(3), 128–128.

Janda, K. and Topouzi, M. (2016). 'Telling tales: using stories to remake energy policy'. *Building Research and Information*, *43*(4), 516–533.

Jochem, E. (Ed.). (2004). *Steps towards a sustainable development: A white book for RandD of energy-efficient technologies*. Zurich, Switzerland: ETHZ Press.

Joller, L. and Urmas, V. (2016). 'Learning from an electromobility living lab: Experiences from the Estonian ELMO programme'. *Case Studies on Transport Policy*, *4*, 57–67.

Juujärvi, S. and Pesso, K. (2013). 'Actor roles in an urban living lab: What can we learn from Suurpelto, Finland?' *Technology Innovation Management Review*, *3*(11), 22–27.

Karvonen, A., Evans, J. and van Heur, B. (2014). 'The politics of urban experiments: Radical change or business as usual?' In M. Hodson and S. Marvin (Eds.), *After sustainable cities*. London: Routledge.

Karvonen, A. and van Heur, B. (2014). 'Urban laboratories: Experiments in reworking cities'. *International Journal of Urban and Regional Research*, *38*(2), 379–392.

Kesselring, P. and Winter, C. J. (1994). 'World Energy Scenarios: A Two-kilowatt Society, Plausible Future or Illusion?' Paper presented at the Proceedings of the Energietage, Villigen, Switzerland.

Kharrazi, A., Rovenskaya, E., Fath, B. D., Yarime, M. and Kraines, S. (2013). 'Quantifying the sustainability of economic resource networks: An ecological information-based approach'. *Ecological Economics*, *90*, 177–186.

Klein, J. T., Grossenbacher-Mansuy, W., Haberli, R., Bill, A., Scholz, R. and Welti, M. (2001). *Transdisciplinarity: Joint problem solving among science, technology, and society. An effective way for managing complexity*. Basel, Switzerland: Birkhauser.

Lienin, S., Kasemir, B. and Stulz, R. (2004). *Bridging science with application for sustainability: Private–public partnerships in the Novatlantis pilot-region of Basel*. Duebendorf, Switzerland: Novatlantis. Retrieved from www.hks.harvard.edu/sustsci/ists/docs/ists_novatlantis04. pdf (accessed 24 December 2016).

Lienin, S., Kasemir, B., Stulz, R. and Wokaun, A. (2005). 'Partnerships for sustainable mobility: The pilot region of Basel'. *Environment*, *47*(3), 22–35.

Loorbach, D. (2014). *To Transition! Governance Panarchy in the New Transformation*. Rotterdam, The Netherlands: Dutch Research Institute for Transitions.

Marechal, F., Favrat, D. and Jochem, E. (2005). 'Energy in the perspective of the sustainable development: The 2000 W society challenge'. *Resources, Conservation and Recycling*, *44*(3), 245–262.

McCormick, K. and Kiss, B. (2015). 'Learning through renovations for urban sustainability: The case of the Malmö Innovation Platform'. *Current Opinion in Environmental Sustainability*, *16*, 44–50.

Morosini, M. (2010). 'A "2000-watt society" in 2050: A realistic vision?' In M. Mascia and L. Mariani (Eds.), *Ethics and climate change: Scenarios for justice and sustainability*. Padova, Italy: CLEUP.

Mulder, I. (2012). 'Living labbing the Rotterdam way: Co-creation as an enabler for urban innovation'. *Technology Innovation Management Review*, 2(9), 39–43.

Müller, K. and Steinke, G. (2015). 'Transformation of a historical coal bunker into a solar power station using multi-colored BIPV'. Paper presented at the Conference proceedings of the 10th Energy Forum 2015 on Advanced Building Skins, Bern, Switzerland.

Notter, D. A., Meyer, R. and Althaus, H.-J. (2013). 'The western lifestyle and its long way to sustainability'. *Environmental Science and Technology*, 47(9), 4014–4021.

Sengers, F., Berkhout, F., Wieczorek, A. J. and Raven, R. (2016). 'Experimenting in the city: Unpacking notions of experimentation for sustainability'. In J. Evans, A. Karvonen, and R. Raven (Eds.), *The experimental city*. Oxford, UK, and New York: Routledge.

SIA. (2011). *SIA energy efficiency path*. Zurich, Switzerland: SIA.

Stulz, R., Tanner, S. and Sigg, R. (2011). 'Swiss 2000-Watt society: A sustainable energy vision for the future'. In F. Sioshansi (Ed.), *Energy, sustainability and the environment*. Boston, MA: Butterworth-Heinemann.

Trencher, G., Castán Broto, V., Takagi, T., Sprigings, Z., Nishida, Y., and Yarime, M. (2016). 'Innovative policy practices to advance building energy efficiency and retrofitting: Approaches, impacts and challenges in ten C40 cities'. *Environmental Science and Policy*, 66, 353–365.

Veeckman, C. and van der Graaf, S. (2015). 'The city as living laboratory: Empowering Citizens with the citadel toolkit'. *Technology Innovation Management Review*, 5(3), 6–17.

Voytenko, Y., McCormick, K., Evans, J. and Schliwa, G. (2016). 'Urban living labs for sustainability and low carbon cities in Europe: Towards a research agenda'. *Journal of Cleaner Production*, 123(1), 45–54.

11

AGENCY, SPACE AND PARTNERSHIPS

Exploring key dimensions of urban living labs in Vancouver, Canada

Sarah Burch, Alexandra Graham and Carrie Mitchell

1. Introduction

The Paris Agreement to combat global climate change was adopted at the 21st Conference of the Parties (COPs) to the United Nations Framework Convention on Climate Change (UNFCCC) on 12 December 2015 and signed by 177 nations on 22 April 2016. This agreement added another dose of urgency to the climate change discourse by suggesting that warming be kept well below 2°C. But the agreement suffers from the same challenges that plague any international law: domestic policies and legislation must be developed on the part of every signatory to give the treaty force and effect. In other words, municipal, provincial or state, and national governments have power and jurisdiction over the sources of emissions that contribute to a globally unsustainable pathway, and so are seeking to pursue climate change experiments that are domestically desirable, cost-effective and hold the potential to significantly and rapidly reduce greenhouse gas emissions.

Transitions to more sustainable futures remain illusive, however, particularly in rapidly urbanising areas: climate change action plans proliferate, but it is unclear whether the actions contained within these plans will yield the 80 per cent reduction in emissions by 2050 that is called for by many cities, provinces and states.[1] While scholars are increasingly exploring how cities can transition towards more sustainable futures under a changing climate (Burch, 2010a; Coenen, Benneworth and Truffer, 2012; Markard, Raven and Truffer, 2012), questions remain about how to govern this transition, which actors must participate in it, and how successful experiments can be shared with jurisdictions that are lagging behind.

The complexity of the urban climate governance challenge rests not only on the quantity of emissions produced in urban spaces, but is also a function of path dependency of current governance actors and organisational arrangements, relationships (including trust, power differentials and mechanisms for collaboration) among actors, public perceptions of the problem and the human, technical and financial capacity to respond (Burch, 2010b; Burch, Shaw, Dale and Robinson, 2014). As a result, transitions to low-carbon development pathways are deeply social and political endeavours, rather than simple technical fixes. Moreover, these transitions require experiments in reconfiguring the way that municipal government engages with the private sector, civil society, higher levels of government and experts. Such novel governance approaches may complement current city-level action to foster innovation and expedite the transition towards sustainable futures.

New collaborations, known as "urban living labs", are one potential structure to promote processes of imagination, experimentation and innovation. Bulkeley et al. (2015) defined urban living labs as an experiment that fosters learning in a place-explicit (urban) context. Urban living labs typically draw on multiple actors to develop innovative, scalable socio-technical interventions to catalyse sustainable futures. As agencies within cities begin to experiment in this fashion, case-specific evidence is growing to better understand their structure and impact (Baccarne et al., 2014; Castan Broto and Bulkeley, 2013; Nevens, Frantzeskaki, Gorissen and Loorbach, 2013; Voytenko, McCormick, Evans and Schliwa, 2016)

However, Markard et al. (2012) have identified several key gaps in recent explorations of sustainability transitions, of which urban living labs (or climate change experiments in urban spaces) are a part. First, we need to better understand the agency of different actor groups within the experimentation and sustainable transition processes (Garud and Karnøe, 2003; Raven Verbong, Schilpzand and Witkamp, 2011). The ways that actor groups interact, build strategic alliances and exercise power may shed light on how to manage innovation, replicate successful experiments at other scales and contexts and address problematic or dysfunctional relationships. Second, the spatial and institutional context, including multi-scalar dimensions of socio-technical transitions and innovation, is underexplored, leaving a gap in our understanding of the place-based nature of transitions (Coenen and Truffer, 2012). In particular, without a deeply contextualised view, we may neglect characteristics of transitions that would play out differently in lower income cities, with fundamentally different socio-spatial relations and materialities.

This chapter aims to address these key gaps by exploring the governance of climate change experiments that operate within the urban living lab of the City of Vancouver, Canada. It will investigate the agency of different actors at play in these cases, the spatial context, and the multi-scalar dynamics of efforts to trigger innovation on climate change. First, we review recent literature on socio-technical transitions, climate innovation and urban living labs to uncover the key insights into the drivers of innovation and sources of path dependency. Second, we narrow our analysis of the literature to examine the concepts of agency (with a particular focus on sources of leadership and the value of public–private partnerships in sustainability transitions), scale and spatial context. Third, we will present three cases demonstrating different styles of climate innovation within the City of Vancouver. Finally, we will end with a discussion teasing apart the influence of municipal governments and multi-scalar partnerships within climate innovation to ultimately offer potential solutions that might expedite the socio-technical transition to a more sustainable future.

2. Climate change experimentation in urban spaces

Greenhouse gas emissions and vulnerability to climate change impacts are not simply shaped by climate change policy, but rather the underlying, place-based, development path (Burch et al., 2014; Shaw, Burch, Kristensen, Robinson and Dale, 2014). This consists of the materialities and spatial dimensions of the urban landscape (such as the proximity of work to recreation, housing, and transportation options), the values that give rise to varying levels of consumption, and the ecological context of urban spaces (Burch, 2010a; Dusyk., Berkhout, Burch, Coleman and Robinson, 2009). Reaching the ambitious climate change targets set by the international community, and those emerging at the national level in Canada, requires actors to consider transformative rather than incremental actions. Transformation has been described as responses that produce non-linear changes in systems, and can be triggered by addressing the "root causes" of unsustainable development pathways (such as dysfunctional social or economic arrangements) rather than the "symptoms" (dwelling quality, vehicle efficiency, etc.) (Pelling, O'Brien and Matyas, 2015). Exploring synergies between climate change adaptation, mitigation and other sustainability priorities (such as biodiversity and social equity, for instance) (Beg et al., 2002; Burch et al., 2014; Shaw et al., 2014) may help to yield these transformative outcomes.

The task for both scholarship and practice then becomes one of deepening our understanding of the "seeds" of socio-technical transformations in communities. What are the conditions under which "radical novelties" (after Geels, 2002; Geels and Schot, 2007) emerge, and how can solutions to sustainability challenges be scaled up, expanded and shared? The domains of socio-technical systems theory, paired with an understanding of multi-level governance, begin to shed light on how these novelties emerge.

2.1. Agency, scale and spatial context in urban sustainability transitions

Technological innovations do not occur in a vacuum, but rather are deeply embedded in social practices, cultural values, governance structures and societal rules. This embeddedness may create path dependency or inertia behind unsustainable behaviours or technologies with implications for intentional efforts to engender system transformation. The study of these "socio-technical systems" has gained momentum over the last 15 years, leading to four main conceptual approaches: transition management (employing complex adaptive systems theory and governance insights to provide a practice-oriented model of how to influence change), strategic niche management (the deliberate creation of protected spaces within which a radical novelty might grow to destabilise the dominant rules and structures), multi-level perspective (exploring the dynamic interplay among niche innovations, rule regimes and the broader "landscape")[2] and technological innovation systems (focused on how innovative technologies emerge, and the institutional or organisational changes that accompany this) (Markard et al., 2012). Repeated calls have been made for transitions studies to more deeply explore socio-spatial dynamics (why does place matter, and how does society operate in place?), questions of agency (who leads change under what conditions) and the overlooked multiscalar dimensions of niches and regimes (Law, 2004; Murdoch, 1998).

Sub-national and non-state actors are playing an important role in transformative change as power and responsibility is shifting away from the centralised, nation-state in some countries (Bulkeley and Betsill, 2005; Hooghe and Marks, 2001). While this may be viewed as a "neoliberal turn" that also has negative implications for justice, representativeness and equity (Brenner and Theodore, 2002), it also shifts our attention to theorising governance to include a shifting, overlapping, constellation of actors, variously endowed with capacity (human, social, financial and technical) and jurisdictional authority. Municipal governments, for instance, are well positioned to be active participants in the multi-level governance of sustainability transitions, but their ability to develop and implement innovative sustainability solutions may be impaired by institutional inertia, jurisdictional constraints and limited opportunities to learn from both success and failure in other communities. Depending on the national political structure within which they operate, cities can be "the operational units in which concrete actions can be envisaged, designed, (politically) facilitated and effectively rolled out" (Nevens et al., 2013, p. 111); "cities can even be considered as potential 'motors' for sustainable development (Rotmans et al., 2000) or 'hubs' for extreme innovation" (Bulkeley and Castan Broto, 2012; Ernstson et al., 2010a).

Cities also provide geographical proximity in which local decision makers, entrepreneurs, citizens and consumers coalesce to create natural testing grounds for new ideas and approaches to addressing problems arising from human–environment interactions (Schroeder et al., 2013). The effect of city-scale experiments may not be limited to this rather modest geographic space, however. According to the theory of boomerang federalism, local initiatives may stimulate federal policies that then in turn come back to support the local environment (Fisher, 2013). This parallels the insight described above by sustainability transition theory, in which niche innovations build momentum and put pressure on regimes.

Questions remain, however, with regard to the role of urban actors in triggering and accelerating transformation. While municipal governments have a critical role in experimentation they often act alongside other actors and in a variety of partnership modalities (Castan Broto and Bulkeley, 2013). Experimentation is one tool to open up new political spaces for governing climate change in cities, and draws attention to climate change action that occurs outside of the traditional formalised policy realm (Castan Broto and Bulkeley, 2013). NGOs, corporations, international organisations and community groups, for instance, play a role in global and local environmental governance (Bulkeley and Newell, 2010).

Of the 627 experiments across 100 cities reviewed by Bulkeley and Castan Broto, for instance (Bulkeley and Castan Broto, 2013; Castan Broto and Bulkeley, 2013), 48 per cent involved partnerships. Local governments led most partnerships (in 59 per cent of cases), followed by private sector leaders. Most private-led projects focused on urban infrastructure and most community-led projects focused on the built environment. They conclude that the growing number of project-based experiments should not be viewed as "spill-over effects of a governance system lacking capacity" (p. 9), but as a central component of transition governance since cities are usually approaching sustainability on a case-by-case basis, depending on funding and windows of opportunity.

The agency of both small- and medium-sized enterprises (SMEs) and civil-society groups also appears to be particularly central to urban sustainability experiments and transitions. Burch et al. (2013), for instance, argues that SMEs have untapped potential for innovation as they seek ways to reduce costs, win new customers, receive good publicity and have higher staff retention. Community groups have the opportunity and motivation to innovate in niche spaces, and a successful example presented was Portland's Solarise scheme (Aylett, 2013), highlighting how "civil society and local businesses can act with a speed and tolerance for risk that is far higher than that of municipal governments" (p. 765). While the majority of SMEs do not position themselves as transformative, a small subset of frontrunner SMEs is actively engaged in introducing radical change to existing market systems. Typically managed by "post-conventional leaders" (Boiral et al., 2014), these sustainability entrepreneurs are often distinguished by an ideological dismissal of prevailing institutional structures like the motivation for profitmaking. Viewing their businesses as agents of change and an extension of their own ideals rather than chiefly as agents of commerce provides these SMEs the freedom to experiment with more radical innovations (Parrish and Foxon, 2008). These frontrunner SMEs reconcile and equate sustainability obligations with corporate strategy, rather than viewing them as competing priorities (Beveridge and Guy, 2005).

In sum, socio-technical transitions theory sheds light on the ways in which sustainability innovations emerge, spread and may come to challenge incumbent technologies or social practices. The urban is a compelling context within which to explore the multi-level governance of innovation, but more clarity is needed regarding the roles of (and interactions among) various actors, as well as the spatial, and multi-scalar nature, of transitions.

2.2. The urban living lab: A conceptual tool for exploring the multi-scalar, multi-actor reality of climate change experiments in cities

Urban living labs are considered a form of experiment that fosters learning in a place-explicit (urban) context with multiple actors to develop innovative, scalable, socio-technical interventions to transition to a more sustainable futures

(Bulkeley et al., 2015). Key characteristics of ULL identified include geographical embeddedness, experimentation and learning, participation and user involvement, leadership and ownership and evaluation and refinement (Voytenko et al., 2016). Urban living labs are a novel approach that may complement current city-action to foster innovation, potentially expediting the transition towards sustainable futures and are emerging as a form of collective urban governance that may address some of the challenges associated with path dependency, distributed authority and varying legitimacy identified by others (see above).

The territorial nature of living labs has been stressed, as well as the importance of public–private partnerships. For instance, "user communities" can be viewed as co-creators instead of traditional observed subjects or stakeholders in experiments (Nevens et al., 2013), a perspective that is central to a more participatory brand of decision-making that includes multiple types and sources of knowledge (cf. Nyong et al., 2007; Olsson and Folke, 2001) and explicitly addresses the normative dimensions of sustainability (Swart et al., 2004).

The City of Vancouver is an experiment in urban sustainability: by aiming to be the "greenest city in the world by 2020", the city has directly significant resources towards progress on clean water, green buildings, local foods, transportation and a green economy. This has spurred an array of smaller-scale initiatives, three of which are the subject of this chapter. In the sections that follow, we apply the concept of an urban living lab to the case of the City of Vancouver: a municipality that has set the stage for a variety of sustainability experiments to take place, with potentially transformative results (Shaw et al., 2014).

3. Socio-technical climate innovators

The goal of this study was to investigate the agency of different urban living lab actors, as well as the spatial context and multi-scalar dimensions of these socio-technical climate innovators. Projects within the City of Vancouver were purposely selected (Creswell, 2014) based on their mission to promote place-explicit experimentation as a means to foster scalable interventions that generate more sustainable urban futures (Bulkeley et al., 2015). Three projects were identified through consultation with local practitioners and academics: (1) CityStudio, (2) Green and Digital Demonstration Platform and (3) Climate Smart. Table 1 outlines the key characteristics of the three cases.

Data were gathered in 2016 from internally published reports, press releases, internal memoranda, official website documents and end of year reports. Content analysis was used to investigate the overarching project goals, the agency and power among actors, the impact of current practices and the scale of operation. Three semi-structured interviews with senior staff at each organisation were carried out to triangulate findings from the content analysis, and gather additional documents that were unavailable online.

The three projects examined here demonstrate the different governance styles, agency dynamics and spatial dimensions of urban living labs. These three projects

TABLE 11.1 Key characteristics of three Vancouver-based projects that illustrate urban sustainability experimentation

	CityStudio	Green and Digital Demonstration Program (GDDP)	Climate Smart
Start date	2009	2014	2007
Program goal(s)	(1) Innovate and prototype solutions to complex sustainability problems (2) Create a culture of change at City Hall (3) Demonstrating possibilities to residents (4) Empowering students to be change makers	(1) Accelerate sustainable and digital innovation (2) Commercialize green innovation (3) Generate green jobs	(1) Foster change that reduces greenhouse gases within established small and medium businesses
Process	Design prototypes using five Ss: staff, support, site, steward, scale	Provide city assets for product demonstration and testing for up to two years to selected SMEs	Provide one-on-one training, a unique web-based software that tracks emissions, and peer-to-peer learning
Actors	(1) City of Vancouver staff (2) CityStudio staff (3) Post-secondary professors (4) Post-secondary students (5) Community members (6) Leadership council (university presidents + City manager) (7) Operations council (university and City staff)	(1) Vancouver Economic Commission Staff (2) City of Vancouver Staff (3) GDDP participants (SMEs)	(1) SMEs (2) Municipal staff (3) Provincial staff (4) Federal staff (5) Climate Smart staff
Scale	City scale	City scale	City, regional, provincial and national scale

continued . . .

TABLE 11.1 Continued

	CityStudio	*Green and Digital Demonstration Program (GDDP)*	*Climate Smart*
Impact	Program engaged 3500 students, 163 faculty members, 75 City of Vancouver staff, 119 community advisors and 195 projects on the ground	30 companies applied and approximately ten were selected to participate in year one	Climate Smart has worked with over 800 business, 30 host regions (Climate Smart, 2016), and annually manages more than one million tonnes of carbon dioxide

represent different phases of innovation and engage with different actors: CityStudio focuses on engagement, culture shift and prototyping; VEC focuses on supporting implementation and commercialisation and Climate Smart fosters change within established small and medium businesses. Together the projects demonstrate a range of leading actors and socio-spatial contexts.

3.1. CityStudio Vancouver

CityStudio Vancouver is an innovation hub embedded within the City of Vancouver's municipal government. The program engages university students from six local post-secondary institutions, City staff and community members to co-design and execute high-impact projects. CityStudio's mission is to help make better cities, where "City Halls get ideas, energy and hard work from students in exchange for the practical experience and skills that will forge future leaders" (CityStudio, 2015). Projects initially focused on the broad notion of sustainability and have now expanded to include healthy, liveable city targets.

CityStudio launched in 2009, after co-founders Duane Elverum and Janet Moore won the City of Vancouver ideas competition during the City's greenest city citizen engagement process (Elverum and Moore 2015, November 14). Now it its fourth year, CityStudio has engaged 3500 students, 163 faculty members, 75 City of Vancouver staff, 119 community advisors and 195 projects on the ground (CityStudio, 2015). CityStudio has several initiatives to promote innovation and citizen engagement: a one-semester intensive studio program for undergraduate students, a network of courses across six campuses that incorporate CityStudio deliverables, a one-year master's studio program, and a consulting service for other cities to launch their own CityStudios.

The experimentation processes nurtured at CityStudio focuses on empowering students with the tools to be actively engaged in the city building process. Co-Director, Duane Elverum likened this model to a teaching hospital: "Imagine the teaching hospital model applied to City Hall, opening its doors to students for hands-on learning" (Elverum and Moore, 2014). The CityStudio program incorporates the latest research in innovation, group dynamics, and behavioural change (D. Elverum, personal communication, 12 April 2016) to prototype urban innovations. The key dimensions of the projects include:

1   Staff: Build intentional relationships with City staff who have decision-making power, time, and funding.
2   Support: Write the project story to connect with people and generate funding and materials.
3   Site: Launch the project in a real site to demonstrate the tangible benefits.
4   Steward: Seek project stewards that live nearby and/or are interested.
5   Scale: Design the experiment so it can be scaled and repeated (CityStudio, 2015).

In the one-semester CityStudio program, students apply to participate in the course through Simon Fraser University. City Staff bring the students problems, such as "residents use too much water in July and August". In groups students develop skills in dialogue, design, project management and implementation. Each term ends with a publication and an on-the-ground exhibition of idea prototypes. Examples of projects include FoodShare, which diverted unmarketable but healthy produce from grocery stores to community food programs, and Ask Lauren, which provided a Community Concierge to build relationships and resilience among apartment dwellers (CityStudio, 2015). Co-Director, Duane Elverum, states it is the prototyping process that is unique from other forms of labs and the cornerstone of behavioural change as it evokes a healthy peer pressure for change.

While collaborative models of social innovation are becoming more popular, they are not without their challenges. Since CityStudio emerged out of an idea competition, they were given a neutral office space to begin operating before they even started to develop their governance model. Currently, CityStudio operates under one university, Simon Fraser University, but is considered a collaborative project between six universities and the City of Vancouver. They have two councils that offer strategic advice: The Leadership Council and the Operations Council. The Leadership Council consists of school presidents and a City manager; they meet once a year to receive high-level updates and offer guidance for program development. The Operations Council is made up of three to four staff members at each institution who are influential leaders, progressive thinkers and "slightly disruptive"; the Operations Council meet once a semester to offer more hands-on guidance (D. Elverum, personal communication, 12 April 2016).

What has remained imperative for CityStudio Co-Directors is that they remain the only individuals who have direct authority over operations. They recognise that there needs to be neutral leaders who are "devoted and crazy" to their mission (D. Elverum, personal communication, 12 April 2016). However, they acknowledge that building trust and ownership among collaborative partners is an on-going learning process that they are still developing.

Going forward, CityStudio is planning to move towards a hybrid non-profit organisation model that has a social enterprise division. They believe that the best governance model would be to bring on a general manager, while the founders move towards leadership positions on the board. They believe the future board should be made up of key community stakeholders instead of staff from the six universities and City because it may be challenging for them to take off their "institutional hat" (D. Elverum, personal communication, 12 April 2016).

CityStudio interacts with several spatial scales. Locally, they aim to create deeply engaging university experiences, where students across Vancouver graduate feeling empowered to participate in city building. The program is part of a global shift in education that provides more hands-on learning experiments to help students change the world. Within the City of Vancouver, CityStudio wants to benefit citizens be making the city more sustainable, liveable and joyful by building prototypes that advance city goals.

CityStudio's aim is to create a culture of creativity and innovation within City Hall; this program gives City staff "permission to innovate" (Elverum, personal communication, 12 April 2016). Gregor Robertson, Mayor of Vancouver, states that this program is "energizing our city and our staff, and creating a culture change inside City Hall by encouraging staff to work across boundaries with energy and creativity" (CityStudio, 2015, p. 5).

CityStudio received a grant last year to hold a national conference to share the CityStudio model with cities across Canada, including Brantford, Calgary, Edmonton, Hamilton, Surrey, Toronto, Waterloo and Winnipeg. Going forward, with funding from the McConnell Foundation, they will be working on perfecting the operational model so they can easily share this program with other interested cities.

3.2. Green and digital demonstration program

The Vancouver Economic Commission (VEC), in partnership with the City of Vancouver, launched the urban living lab Green and Digital Demonstration Program (GDDP) in Fall 2014 to support green and digital start-ups (J. McPherson, personal communication, 12 April 2016). The program goals are to accelerate sustainable and digital innovation, commercialisation and job growth (VEC, n.d. b).

The program was inspired by work being conducted at the University of British Columbia, living labs research, and a previous economic development strategy at VEC (J. McPherson, personal communication, 12 April 2016). GDDP supports selected start-up companies by providing city assets for product demonstration and testing for up to two years. Since the City of Vancouver manages approximately $6 billion in assets – including over 500 buildings, 2300 km of roads and bike lanes, 1800 vehicles, a data centre, water and waste utilities and 1300 ha of park space (VEC, n.d. b) – there are numerous opportunities for experimentation. For example, TSO Logic, a software solutions company, connected their new digital dashboard software to the City's computer servers to demonstrate the inefficiency of the City's data centre operations.

Interested companies are encouraged to apply the GDDP program and are evaluated based on the following key criteria: scalable, environmental, implementable, minimal risk, and no direct cost to the City (VEC, n.d. a). Scalability is critical to ensure future sales growth and local jobs. The environmental impact must be direct and measurable. Products and services must be market-ready. The product must score a 7–9 on the technology readiness level scale (J. McPherson, personal communication, 12 April 2016), which means the prototype must be at, or near, the planned operational use. Risk has to be removed from the City and all technology has to be vetted. Finally, the City cannot be responsible for any direct or incremental costs (VEC, n.d. a). As previously mentioned, municipal governments have limited opportunities to learn from experimentation, since failure has significant financial and political risks associated with it. However, failure is a critical ingredient to innovation. Since governments are not incentivised to take risks, this third-party led innovation can provide the benefits of innovation to the city while reducing associated liabilities.

Based on the aforementioned criteria, VEC presents a shortlist of applications to the GDDP steering committee, made up of senior city managers across departments. Shortlisted applicants will pitch to the committee in Dragon's Den fashion, where committee members ask questions to evaluate potential and risk (J. McPherson, personal communication, 12 April 2016). In the first year of program, 30 companies applied and approximately 10 companies were selected to participate in the program (J. McPherson, personal communication, 12 April 2016).

There are three main participant groups in GDDP: the GDDP participant, the VEC and the City of Vancouver. The selected GDDP participants are responsible for funding all direct and indirect project costs and track and submit metrics of success – including environmental benefits. Most companies involved have less than 30 staff members; however, VEC tries to engage larger companies to showcase new products in Vancouver (J. McPherson, personal communication, 12 April 2016). While participating in the GDDP program does not guarantee future City procurement, participants benefit from staff feedback and improved marketing. One company involved with GDDP, for example was able to raise an additional $3,000,000 from funders because of their new relationship with the City (J. McPherson, personal communication, 12 April 2016).

The VEC acts as a project manager and liaison between the City and the GDDP participant. Before this program, companies would connect with City departments on an ad-hoc basis to pilot projects and services (J. McPherson, personal communication, 12 April 2016). Having GDDP has helped streamline the process and ensure selected participants meet a robust set of criteria. This intermediary role is similar to CityStudio's role of streamlining and improving a process to connect keen students with City staff who need additional resources. The VEC is also responsible for the screening process, monitoring and reporting of environmental and job-related achievements, and promotion of successful projects (VEC, n.d. a).

The City approves participants based on feasibility and provides in-kind support, such as feedback and references. This program has helped the City implement its Greenest City Action Plan, which positioned VEC as responsible for creating green jobs and engaging local businesses to green their production.

The GDDP program functions on a highly localised scale within the City of Vancouver. The localised impact of the program includes green jobs and local environmental benefits. However, GDDP aims to accelerate commercialisation of green and digital products and services to be sold both within and beyond city boundaries.

3.3. Climate Smart

Climate Smart is a social enterprise that helps SMEs measure and reduce carbon emissions, cut costs and share their progress externally and internally. They empower companies to innovate through training, a unique web-based software and peer-to-peer learning. Climate Smart launched in 2007 as a pilot program aiming to reduce barriers for SMEs to engage in carbon reduction as part of the non-profit Ecotrust Canada. In Canada, 99.8 per cent of businesses are SMEs, and they employ 64 per cent of the workforce (Innovation Science and Economic Development Canada, 2013). Climate Smart wanted to offer affordable, effective, support to this overlooked business sector.

After initial success in their pilot year, the project incorporated as a social enterprise model by raising socially responsible capital to invest in carbon tracking software (E., Sheehan, personal communication, 14 April 2016). Today they have worked with over 800 business, 30 host regions (Climate Smart, no date), and annually manage more than one million tonnes of carbon dioxide (BC Hydro, 2014) Their program helps reduce an estimated 72,000 tonnes of carbon dioxide each year and helps the business sector save approximately $28 million annually (E., Sheehan, personal communication, 14 April 2016). Other participant benefits include increased staff engagement and reduced turnover (BC Hydro, 2014).

Climate Smart supports business innovation by targeting businesses directly, and also by supporting host partners. Working with businesses, Climate Smart provides a web-based software-as-a-service (Saas) carbon calculator. Climate Smart also provides Client Advisors, who offer direct one-on-one support as businesses experiment and communicate their actions. Additionally, Climate Smart facilitates peer-to-peer learning, which helps accelerate the up-take of innovative ideas. Peer-to-peer activities include membership events and case studies. President Elizabeth Sheehan shared, "We've enjoyed hearing how much peers are learning from each other" (BC Hydro, 2014). For example, one company had success using a cardboard compactor to reduce transportation costs and the idea spread "like wildfire" (BC Hydro, 2014).

Climate Smart also works with host partners, which are often local governments and occasionally large financial or insurance agencies, to help them connect economic development with carbon reduction. For example, Climate Smart teamed up with the City of Vancouver and the VEC to help the neighbourhood False Creek Flats become the greenest place to work in the world (E., Sheehan, personal communication, 14 April 2016). Together, they created an energy and emissions profile for the entire area, identified key business sectors, created a business engagement plan and rolled out a leadership group made up of key businesses.

A key component of their engagement process is their data-driven support. The software has proved a pivotal asset to Climate Smart's success, as it has allowed them to aggregate, analyse and report on carbon data (E., Sheehan, personal communication, 14 April, 2016). Pietra Basilij, at the VEC stated, "we recognize that businesses can be the engine of sustainable community development, but they often need some guidance, and that guidance needs to be data-driven to be effective" (Climate Smart, 2016).

The core actors involved in Climate Smart include businesses, cities and regions, and provincial and federal government actors. Climate Smart provides software and support to businesses and regions, but has no direct control over their actions. Similar to CityStudio and GDDP, Climate Smart acts as an essential lever to accelerate climate innovation. However, they deviate in their spatial scale. Climate Smart's focus is not specifically place-based, like many urban living laboratories. While they work within regions, they also focus extensively on sector-based strategies by working across supply chains, value chains and wider business networks. Using aggregate data from current Climate Smart businesses, they work with several industries including food manufacturing, construction, manufacturing and clean tech, and offices (E., Sheehan, personal communication, 14 April, 2016).

Geographically, Climate Smart mostly operates within the Province of British Columbia, with an aim to expand more across Canada and internationally. They work with the province and federal government to provide feedback and recommend policies that would further incentives SMEs to transition to sustainable, low-carbon practices.

4. Comparative analysis: Commonalities and challenges

As the above results demonstrate, urban living laboratories present a novel opportunity for municipal governments to experiment with innovative tools and mechanisms to transition to more sustainable urban futures. In the section that follows we will examine some of the commonalities between the case studies, namely: (1) the critical role of local government in the function of urban living labs; (2) the importance of place-based problem identification in facilitating sustainability transitions; (3) the value of multi-scalar and multi-institutional relationships to the success of urban experimentation and (4) the vital role of urban living labs play in diverting risk and concentrating rewards for city governments and their constituents. Despite their potential, there are also potential pitfalls of relying on urban living labs to catalyse sustainability transitions in cities, including the implications of outsourcing innovation, the possibility of misaligned mandates and the potential for piecemeal environmentalism. These challenges will be discussed at the end of this section.

4.1. Place-based problem identification and problem solving

Urban living labs, while concerned with societal issues operating at various geographic and institutional scales, are inherently local. Their success, defined by their ability to *demonstrate* and *drive* socio-environmental action in cities, is critical to fostering shifts toward sustainability. To do this effectively, however, requires active participation from a variety of stakeholders who each have a vested interest in catalysing change. Literature from the field of environmental psychology suggests that a positive attachment to place correlates with people's willingness to participate in its protection (Brehm et al., 2006; Buta et al., 2014; Halpenny, 2010; Vaske and Kobrin, 2001), and that place attachment is critical to community participation in planning (Manzo and Perkins, 2006). Our case studies suggest that place attachment may also be an important factor in the success of urban living labs.

CityStudio, for example, relies on City of Vancouver staff to bring real, local, problems to students studying and living in Vancouver, such as "residents use too much water in July and August". Critical to the success of the CityStudio program is the showcasing on-the-ground solutions, further grounding the lab in real, local, problem identification and problem-solving activities. GDDP, in partnership with the City of Vancouver supports start-up companies by allowing them to gain temporary access to city assets, such as buildings, streets, or utilities. Entrepreneurs, as a result, identify and solve local problems, and demonstrate technologies and implement proof-of-concept trails in place, in real time. Climate Smart, while focused on small and medium size businesses and non-profit organisations, also utilises place-based problem identification. In the case of Climate Smart, "place" is highly localised – the individual business or non-profit organisation.

Overall, we find that place matters in terms of problem identification, effectively demonstrating proof-of-concept, and garnering support and participation from various stakeholders – including city staff, local businesses, university students and the wider community. While the aim of some of the work of the ULL examined does involve creating scalable technologies, products or planning interventions, each example is grounded and shaped by place-based politics and support structures. Therefore, innovation can be a product of actors possessing a sense of place (generating different sets of potential actions depending on local identities and geographies), but the *specific* nature of place also matters. Different cities have varying degrees of institutional inertia, distinctive environmental complexities, competing priorities and different sustainability champions.

4.2. Multi-scalar and multi-institutional relationships

Tackling climate change at the urban scale necessarily entails engaging with non-state actors and traditional municipal planning mechanisms (Bulkeley and Betsill, 2013). The experience of urban living labs in Vancouver suggests similar multi-scalar and multi-institutional processes. CityStudio, for example, engaged almost 2,000 students, 75 university faculty members across six campuses, 40 City of Vancouver staff and over 100 other stakeholders in dialogue over a three year time period. City Studio, along with GDDP and Climate Smart have essentially created the literal, figurative and at times virtual, niche space to test radical innovation. However, niches, or the locus for radial innovation, can only metaphorically break existing socio-technical regimes if there is an appetite for change, and windows of opportunity in which to break through (Geels, 2010).

The City of Vancouver enables niche innovation to flourish through productive multi-scalar and multi-institutional relationships. In the case of CityStudio, the City initiated ideas competition in which CityStudio competed and won. Since inception, the City has provided both office space and support staff to CityStudio, along with real world problems to tackle. GDDP also capitalises on strong multi-scalar and multi-institutional relationships through access to City assets and supportive provincial policies. Essentially, the City has unlocked and destabilised their own regime, creating windows of opportunities in which innovation can flourish. Climate Smart uses global sustainability standards to help SMEs and non-profits calculate their businesses' carbon footprint and measure emissions. Furthermore, they utilize global carbon offset markets to help businesses reduce their emissions. Again, we see multi-scalar and multi-institutional relationships at play. Progressive policies at the municipal and provincial levels of government catalyse businesses innovation efforts, enabling change.

Overall, we find the three case studies from Vancouver demonstrate the importance of multilevel (including various geographic, political and institutional scales) relationships. Without positive, enabling interactions across scales of influence these urban living labs would be minimal. However, it is unclear from this analysis whether the enabling environment in which these ULL operate has led to regime shift. While Mayor Robertson claims, for example that City Studio is "creating a culture change inside City Hall", we do not have the evidence to claim that there is a causal link between niche innovation in Vancouver and regime transitions. These cases suggest that multi-level and multi-scalar innovation could be effective because they create boundary products that are accountable to multiple actors. For example, in the GDDP case, the City is the targeted end-user for the innovations being tested; by involving the City as a leading partner with a moderate level of influence but a minimised level of risk, the City might be more likely to adopt the product in the future. This suggests that the multi-level process has the potential to change the institutions participating in the ULL. Additionally, a multi-level process has the potential for success because it enables replication in actors across geographic scales, as demonstrated by the uptake of the CityStudio model across Canada. Further research is warranted to study this connection in greater detail.

4.3. Redistributing agency and diverting risk

The final similarity among the three case studies examined in this chapter centres around the issue of agency. In these three cases each organisation is able to varying degrees, act independently, but simultaneously be supported by municipal and/or provincial governments. This degree of independence runs counter to traditional urban planning theory, in which power and agency is vested with planning professionals and within municipal bureaucracies. This "rational-comprehensive" approach is highly structured, and, for better or worse, has been the basis of Western planning theory and practice for decades. Urban living labs flip traditional planning theory and practice, redistributing agency to non-traditional urban actors.

CityStudio, for example emerged out of the City's greenest city citizen engagement process, and while the organisation is physically embedded in City Hall it was developed, and is currently led by, people outside of municipal government. City staff proposes place-based problems for students to work on, but the agency to innovate is left in the hands of students and mentors. Similarly, GDDP provides access to government assets, but the agency to develop solutions is placed with entrepreneurs, rather than city councillors, planners or city constituents. Climate Smart transfers agency to SMEs and non-profit businesses by way of evaluating areas for potential improvement in environmental performance.

The end result for municipal government is that most of the financial and institutional risk associated with innovation is shared, or partially redirected, to private organisations and citizens. Moreover, rather than using staff time and municipal resources to test new planning ideas, ULL take responsibility for prototypes and proof-of-concept testing. This has obvious advantages for cities, as they can capitalise on new technologies, tools and projects after they have been tested in situ. There are, however, critical questions related to this redistributing agency and risk.

4.4. Potential pitfalls

While urban living labs offer great potential to facilitate the breakthrough of niche innovation into wider socio-technical regimes, they may also cause challenges for successful sustainability transitions. The first challenge relates to agency and risk redistribution. While there are obvious advantages to redistributing the risk of innovation for cities from a financial and administrative perspective, the drawback of outsourcing innovation is that internal agency may be diminished.

Moreover, without strong internal leadership and direction, it is possible that niche innovations could co-opt cities' master planning activities and redirect cities' focus and planning in unexpected ways. Contemporary approaches to environmental management have often been piecemeal, particularly in the business sector (Welford, 2013); outsourcing innovation to ULL could lead, without proper municipal oversight, to sporadic urban sustainability efforts. Whether these efforts, and any resulting shifts in the pathways that cities follow toward sustainability, are ultimately desirable and effective will likely vary widely from city to city.

We should not assume that the mandate of ULL and cities are congruent; a focus on niche innovation and environmental entrepreneurism may come at the expense of social and environmental planning in cities. While et al. (2004, p. 552) grapple with what they refer to as the "sustainability fix", or the "[s]elective incorporation of environmental goals, determined by the balance of pressure for and against environmental policy within and across the city". In the United Kingdom, they found that environmental policy was largely driven by an economic rationality that mostly restricted environmental strategies to "win–win" style interventions that were "either non-threatening or complementary to prevailing growth strategies" (566). In the case of GDDP in Vancouver, for example innovations "must be proven cost-neutral or feasible before the city will implement"; a similar win–win imperative is inherent in the design of Climate Smart, whereby clients can "save money" by "simply turning off lights" (website). This "sustainability fix" is far from the "radical innovations" envisioned in sustainability transitions literature.

5. Conclusion

Transitioning towards sustainable futures is a process of experimentation, fraught with novelty, failure and opportunity. Cities are well positioned to be active participants in the multi-level governance of these transitions, but their ability to develop and implement innovative sustainability solutions may be impaired by institutional inertia, jurisdictional constraints and limited opportunities to learn from both success and failure in other communities. New collaborations, known as "living labs", have been proposed as complements to city-led action that establish niche conditions for social and technical innovation. These arrangements may have transformative potential because of their higher tolerance for risk, multi-actor engagement, and faster pace of action (Bulkeley et al., 2015). As these cases show, however, multi-actor engagement and rapid action may even occur in cases where tolerance for risk is quite low.

The cases described here suggest that it is valuable to empower agents who act as intermediaries between innovators and the municipal government. Two of the three cases created a streamlined a process of support for these agents that otherwise was chaotic and unproductive for the City. The cases varied widely, however, in the quantity and breadth of individuals who were the targets of engagement processes, suggesting that innovation processes may range from exclusive to inclusive, depending on the end goal of the initiative. CityStudio's goal was to empower all students to become change-makers and city builders, regardless of financial resources and experience. Climate Smart was similarly inclusive and aimed to reduce emissions in SMEs by focusing on both the economic and environmental benefits. GDDP, however, was more exclusive in nature by having a competitive selection process to ensure projects were scalable, environmental, and implementable with minimal risks.

Complex patterns of actor interactions emerged in these cases, illustrating that some actor groups (such as small businesses) often face coordination challenges and human, technical, and financial capacity gaps. As such, partnership models that directly address these challenges might be more successful in stimulating accelerated innovation. Ultimately, it is clear that urban living labs are deeply place-based phenomena, with significant (often untapped) potential to produce transformative shifts toward more sustainable development pathways in urban spaces.

Notes

1 See, for example the British Columbia climate change action plan, Ontario's Five-year Climate Change Action Plan, and the City of Vancouver's climate change plan (City of Vancouver, 2012; Province of British Columbia, 2008; Province of Ontario, 2016).
2 Niches are viewed as protected spaces within which low-carbon (or other) innovations might emerge, regimes as interlocking systems of rules that are characterised by path dependency (but might be destabilised by niche-level innovations), and landscapes as exogenous factors such as socio-political shifts and the macro-economic context (Geels et al., 2016).

References

Aylett, A. (2013). 'Networked urban climate governance: Neighborhood-scale residential solar energy systems and the example of Solarize Portland'. *Environment and Planning C: Government and Policy, 31*, 858–875.

Baccarne, B., Schuurman, D., Mechant, P. and De Marez, L. (2014). XXV ISPIM Innovation Conference, Proceedings, (2014) The role of urban living labs in a smart city, 25th ISPIM Innovation Conference, Dublin, Ireland.

BC Hydro. (2014). Climate Smart helps 775 companies find the business case for energy efficiency. Retrieved from www.bchydro.com/news/conservation/2014/climate-smart. html?WT.mc_id=b-14–01_climate-smart

Beg, N., Morlot, J.C., Davidson, O., Afrane-Okesse, Y., Tyani, L., Denton, F., . . . Atiq Rahman, A. (2002). 'Linkages between climate change and sustainable development'. *Climate Policy, 2*, 129.

Beveridge, R. and Guy, S. (2005). 'The rise of the eco-preneur and the messy world of environmental innovation'. *Local Environment: The International Journal of Justice and Sustainability, 10*, 665–676.

Boiral, O., Baron, C. and Gunnlaugson, O. (2014). 'Environmental leadership and consciousness development: A case study among Canadian SMEs'. *Journal of Business Ethics, 123*, 363–383.

Brehm, J. M., Eisenhauer, B. W. and Krannich, R. S. (2006). 'Community attachments as predictors of local environmental concern: The case for multiple dimensions of attachment'. *American Behavioral Scientist, 50*(2), 142–165.

Brenner, N. and Theodore, N. (2002). 'Cities and the geographies of "actually existing neoliberalism"'. *Antipode, 34*, 349–379.

Bulkeley, H. and Betsill, M. (2005). 'Rethinking sustainable cities: Multi-level governance and the "urban" politics of climate change'. *Environmental Politics, 14*, 42–63.

Bulkeley, H. and Betsill, M. M. (2013). 'Revisiting the urban politics of climate change'. *Environmental Politics, 22*(1), 136–154.

Bulkeley, H., Breitfuss, M., Coenen, L., Frantzeskaki, N., Fuenfschilling, L., Grillitsch, M., . . . Voytenko, Y., (2015). *Theoretical framework working paper on urban living labs and urban sustainability transitions*. Lund, Sweden: Lund University.

Bulkeley, H. and Castan Broto, V. (2013). 'Government by experiment? Global cities and the governing of climate change'. *Transactions of the Institute of British Geographers, 38*, 361–375.

Bulkeley, H. and Newell, P. (2010). Governing climate change. London: Routledge.

Burch, S. (2010a). 'In pursuit of resilient, low-carbon communities: An examination of barriers to action in three Canadian cities'. *Energy Policy, 38*, 7575–7585.

Burch, S. (2010b). 'Transforming barriers into enablers of action on climate change: Insights from three case studies in British Columbia, Canada'. *Global Environmental Change, 20*, 287–297.

Burch, S., Schroeder, H., Rayner, S. and Wilson, J. (2013). 'Novel multi-sector networks and entrepreneurship in Metro Vancouver: A study of small business as an emerging non-state actor on climate change mitigation'. *Environment and Planning C, 31*, 822–840.

Burch, S., Shaw, A., Dale, A. and Robinson, J. (2014). 'Triggering transformative change: A development path approach to climate change response in communities'. *Climate Policy, 14*, 467–487.

Buta, N., Holland, S. M. and Kaplanidou, K. (2014). 'Local communities and protected areas: The mediating role of place attachment for pro-environmental civic engagement'. *Journal of Outdoor Recreation and Tourism, 5*, 1–10.

Castan Broto, V. and Bulkeley, H. (2013). 'A survey of urban climate change experiments in 100 cities'. *Global Environmental Change, 23*, 92–102.

City of Vancouver. (2012). Greenest City 2020 action plan. City of Vancouver, Vancouver.

CityStudio. (2015). *CityStudio Vancouver Year 4 annual report*. Retrieved from http://citystudio vancouver.com/publications/

Climate Smart. (2016). CS brief. *New climate partnerships*. Retrieved from bit.ly/New ClimatePartnerships

Coenen, L., Benneworth, P. and Truffer, B. (2012). 'Toward a spatial perspective on sustainability transitions'. *Research Policy, 41*, 968–979.

Coenen, L. and Truffer, B. (2012). 'Places and spaces of sustainability transitions: Geographical contributions to an emerging research and policy field'. *European Planning Studies, 20*, 367–374.

Creswell, J. (2014). *Research design: Qualitative, quantitative, and mixed methods approaches*. Thousand Oaks, CA: Sage.

Dusyk, N., Berkhout, T., Burch, S., Coleman, S. and Robinson, J. (2009). 'Transformative energy efficiency and conservation: A sustainable development path approach in British Columbia'. *Energy Efficiency, 2*, 387–400.

Elverum, D. and Moore, J. (2014, February 26). 'How to make the city a classroom and free the students'. Globe and Mail.

Elverum, D. and Moore, J. [TEDx Talks]. (2015, November 14). *Every city in the world needs a CityStudio*. Retrieved from www.youtube.com/watch?v=K2OXT85BPH0

Ernstson, H., van der Leeuw, S., Redman, C., Meffert, D., Davis, G., Alfsen, C. and Elmqvist, T. (2010). 'Urban transitions: On urban resilience and human-dominated ecosystems'. *Ambio, 39*, 531–545.

Fisher, D.R. (2013). 'Understanding the relationship between subnational and national climate change politics in the United States: Toward a theory of boomerang federalism'. *Environment and Planning C Govern Policy, 31*, 769–784.

Garud, R. and Karnøe, P. (2003). 'Bricolage versus breakthrough: distributed and embedded agency in technological entrepreneurship'. *Research Policy, 32*, 277–300.

Geels, F.W. (2002). 'Technological transitions as evolutionary reconfiguration processes: A multi-level perspective and a case-study'. *Research Policy, 31*, 1257–1274.

Geels, F.W. (2010). 'Ontologies, socio-technical transitions (to sustainability), and the multi-level perspective'. *Research policy, 39*(4), 495–510.

Geels, F.W., Berkhout, F. and van Vuuren, D. (2016). 'Bridging analytical approaches for low-carbon transitions'. *Nature Climate Change, 6*.

Geels, F.W. and Schot, J. (2007). 'Typology of sociotechnical transition pathways'. *Research Policy, 36*, 399–417.

Halpenny, E. A. (2010). 'Pro-environmental behaviours and park visitors: The effect of place attachment'. *Journal of Environmental Psychology, 30*(4), 409–421.

Hooghe, L. and Marks, G. (2001). 'Types of multi-level governance'. *European Integration Online Paper* 5. Retrieved from http://eiop.or.at/eiop/texte/2001–2011a.htm

Innovation Science and Economic Development Canada. (2013). Key small business statistics – August 2013. Government of Canada.

Markard, J., Raven, R. and Truffer, B. (2012). 'Sustainability transitions: An emerging field of research and its prospects'. *Research Policy, 41*, 955–967.

Nevens, F., Frantzeskaki, N., Gorissen, L. and Loorbach, D. (2013). 'Urban transition labs: Co-creating transformative action for sustainable cities'. *Journal of Cleaner Production, 50*, 111–122.

Nyong, A., Adesina, F. and Elasha, B.O. (2007). 'The value of indigenous knowledge in climate change mitigation and adaptation strategies in the African Sahel'. *Mitigation and Adaptation Strategies for Global Change, 12*, 787–797.

Olsson, P. and Folke, C. (2001). 'Local ecological knowledge and institutional dynamics for ecosystem management: A study of Lake Racken watershed, Sweden'. *Ecosystems, 4*, 85–104.

Parrish, B.D. and Foxon, T.J. (2008). 'Sustainability entrepreneurship and equitable transitions to a low-carbon economy'. *Greener Management International, 55*, 47–62.

Pelling, M., O'Brien, K. and Matyas, D. (2015). 'Adaptation and transformation'. *Climatic Change, 133*(1), 113–127.

Province of British Columbia. (2008). *Climate action for the 21st century*.Victoria, BC: Author.

Province of Ontario. (2016). 'Ontario's climate change action plan 2016–2020'. Toronto, Canada: Government of Ontario.

Raven, R.P.J.M., Verbong, G.P.J., Schilpzand, W.F. and Witkamp, M.J. (2011). 'Translation mechanisms in socio-technical niches: A case study of Dutch river management'. *Technology Analysis and Strategic Management, 23*, 1063–1078.

Rotmans, J., van Asselt, M. and Vellinga, P. (2000). 'An integrated planning tool for sustainable cities'. *Environmental Impact Assessment Reviews, 20*, 265–276.

Schroeder, H., Burch, S. and Rayner, S. (2013). 'Novel multi-sector networks and entrepreneurship in urban climate change governance'. *Environment and Planning C, 31*, 761–768.

Shaw, A., Burch, S., Kristensen, F., Robinson, J. and Dale, A. (2014). 'Accelerating the sustainability transition: Exploring synergies between adaptation and mitigation in British Columbian communities'. *Global Environmental Change, 25*, 41–51.

Swart, R., Raskin, P. and Robinson, J. (2004). 'The problem of the future: Sustainability science and scenario analysis'. *Global Environment Change, 14*, 137–146.

Vancouver Economic Commission (VEC). (No date-a). Acceptance criteria.

Vancouver Economic Commission (VEC). (No date-b). Available city assets.

Vaske, J. J. and Kobrin, K. C. (2001). 'Place attachment and environmentally responsible behavior'. *Journal of Environmental Education, 32*(4), 16–21.

Voytenko, Y., McCormick, K., Evans, J. and Schliwa, G. (2016). 'Urban living labs for sustainability and low carbon cities in Europe: Towards a research agenda'. *Journal of Cleaner Production, 123*, 45–54.

Welford, R. (2013). *Hijacking environmentalism: Corporate responses to sustainable development.* London: Routledge.

While, A., Jonas, A. E. and Gibbs, D. (2004). 'The environment and the entrepreneurial city: searching for the urban "sustainability; fix"in Manchester and Leeds'. *International Journal of Urban and Regional Research, 28*(3), 549–569.

12

PLACING SUSTAINABILITY IN COMMUNITIES

Emerging urban living labs in China

Lindsay Mai

1. Introduction

Given the different public financing models for urban initiatives in China compared to ones in Europe, "urban living labs" (ULL) have not become a popular terminology for urban managers and policy makers. Yet this chapter seeks to identify and examine some of the emerging forms of urban initiatives in China that resemble the defined characteristics of "urban living labs" (Voytenko et al. 2016). Since these emerging urban initiatives are mostly at their initial implementation stage, the analysis in this chapter focuses mainly on discussing their design and practices, but it serves also to generalise their impacts to examine how they may collectively process changes in urban transitions. Conceptual elements of political power, democratic space and governing approaches are also taken into account, suggested by a recent theoretical review on ULL processes (Bulkeley et al., 2015).

This chapter starts with a descriptive analysis of emerging ULL practices in China. Both smart cities experimentations and some dispersed community-based responses are observed, with the former being demonstrations of technologically focused ULL while the latter are more socially oriented. However, these initiatives have been fitted in different types of urban responses and designed by disparate networks of actors. Therefore, despite the geographical proximity among these observed initiatives, they are not bounded institutionally. This is displaying a situation different from what recent research indicated that bounding ULL institutionally and geographically created spaces to facilitate innovations (Evans and Karvonen, 2010; Voytenko et al., 2016). It is thus pertinent to look into what alternative arrangements can be put in place to facilitate mainstreaming, translating and scaling up of innovations through discrete networks of actors.

The chapter also seeks to find out whether these multilevel actors have employed "new techniques of governance" (Hodson and Marvin, 2007) in practicing ULL as governing urban transitions through experiments, and how these are reshaping the conventional urban management practices in China. The subsequent discussion makes use of the analytical device of differentiating governing modes as self-governing, governing by authority, governing with provision and governing through enabling (Bulkeley and Kern, 2006).

Empirical data were collected through semi-structured interviews (see Appendix), along with field observation guided by initiative leaders in Shenzhen and its mangrove area (10 January 2015), as well as in Nansha development zone in Guangzhou. Data and observations were also drawn from on-going research between 2012 and 2016, but the interviews cited in this chapter were done mainly during fieldtrips in China carried out in 2015 and 2016.

2. Practicing innovations through urban living labs

Three different types of ULL are present in Guangzhou and Shenzhen of Guangdong Province, a megacity area located in southern China. First, in Shenzhen, a community organisation formed by volunteers named Green Source has started various grassroots initiatives with place-based social innovations to tackle urban environmental problems including water pollution, mangrove conservation and waste recycling. Second, civic-type public programmes have been arranged in Guangzhou given its status as Guangdong Province's administration centre, and recently these programmes have been steered towards more community-oriented innovations. Third, strategic investments have been accumulated in Nansha at the edge of Guangzhou, a new urban district with special status endorsed by national government, leading to a series of techno-centric urban experimentations.

2.1. Grassroots activism

Influenced by neighbouring Hong Kong where a different political system and expanded democratic space at societal level are institutionalised, grassroots activism is growing in Shenzhen. With the geographical proximity of a more developed social innovation sector in Hong Kong, newly formed community groups in Shenzhen have received timely support in best-practice experience sharing and technical transfers. The growing grassroots activism is also fused with tacit and professional knowledge held by community leaders, the strong intention of tackling community issues and to co-create a more sustainable urban environment.

This grassroots activism is manifest in initiatives of the Green Source Environmental Volunteers Association of Shenzhen. Distinct from urban experimental initiatives led by governments or major enterprises that tend to focus on technological innovations, a wide range of social innovations have been devised and practiced by the grassroots organisation. The Green Source houses three main strands of activities: local mangrove conservation, community-based environmental education and evidence-based pollution monitoring. The mechanism rendering these activities operates on new ways of engaging with citizens and other grassroots community-level organisations, as well as innovative techniques for lobbying and agenda setting with local authorities while generating self-sufficient finance.

BOX 12.1 SNAPSHOT OF A GRASSROOTS URBAN LIVING LAB IN CHINA

Name: Green Source Programmes
Location: Shenzhen (Futian urban district and coastal mangrove areas)
Geographic boundary: Shenzhen
Lead organisation: Shenzhen Green Source Environmental Protection Volunteers Association
Partners (Types): ENGOs, municipal and district government, community centres, schools, professionals
Running time: 2013–present
Funding sources: a mixture of donations, membership fees, service provision incomes and government subsidies
Focus areas: wetland mangrove conservation, community sustainability and environmental education workshops ("Green Fun Communities"), water pollution monitoring (Citizen Watch; "Green Water, Flowing Deep")
Web presence: www.szhb.org

FIGURE 12.1 Mangrove seedlings were planted by Green Source volunteers and fenced up from waves and pollution of irresponsible waste dumping.

One of Green Source's founding ideas was driven by a cause to conserve mangrove wetland in its native field within the boundary of Shenzhen, prompting actions to be taken locally. Several years before the launch of Green Source, a few leading individuals (local environmental activists) who later founded Green Source in 2013 started to recognise the potential problem endangering local mangrove forests, and they conducted an informal study of local mangroves, including the specific locations, species and risks of mangroves. They were ascertained about the local mangrove forest's biodiversity value and benefits of extending natural protection from coastal erosion. With the launch of Green Source, the organisation started to call on local volunteers to visit and collectively safeguard some mangrove areas, areas that the organisation leader had identified but were not designated as protected area by the municipal government (Interview 1).

In addition to conserving the existing mangrove forests, Green-Source volunteers also experimented incubating mangrove seedlings in coastal sites of manageable scale they identified as suitable for new mangrove plantation (Box 12.1). The purpose of mangrove incubator was to mitigate the gradual loss of existing mangrove species with replacement, drawing on field research evidence collected by volunteers specialised in environmental science. After mangrove seedlings were planted, it was realised that regular patrols were crucial given the constant threats of urban solid wastes being dumped from settlements nearby.

A series of grassroots actions was then initiated from bottom up, motivated by mangrove conservation, as a novel way of community engagement. Most of the participating volunteers were ordinary residents living in Shenzhen urban districts, who travelled from the city centre to the mangrove coast over weekends for initiatives led by community leaders. These bottom-up initiatives began to render alternative solutions to public programmes that had been distantly organised by municipal governments and universities, drawing design intentions closer to implementation.

The Green Source was also established and expanded so as to better coordinate the mobilised volunteers and to sustain local sustainability actions. Geographical proximity of human resources is an important factor for the organisation relocation. Since Green Source was first set up in 2013, it had been sharing the office space with an established community centre in one of the most densely populated urban districts – *Futian* – in Shenzhen. The growing organisation and its expansion of project activities were forcing the organisation leaders to seek appropriate venue for relocation. Most of the volunteers registered with the Green Source were locally based, while the organisational resources had been concentrated in Futian on social surveys and environmental monitoring. Considering these locally embedded inputs, the Green Source decided to look for available space in the same district.

Meanwhile in 2014, Futian District Government launched a new space for district-based nongovernmental organisations (NGOs), by sparing a site of elevated real-estate value from private developers and turning a new building into a "Shenzhen Social Organisation Headquarters Base (Futian District)". With a total floor size of 3500 squared metres, this multi-storey building accommodates administration, engagement activities and innovation incubation of local NGOs. It is a showcase of a planned transformation of government functions in Shenzhen. Seizing this opportunity, the Green Source submitted an application to be a tenant of the new building, and opened their new office in 2015.

The various approaches to community sustainability engagement being deployed by the Green Source represent a transformative process of not only community empowerment, but also divergent governing responses. The latter, as denoted by the organisation's chairman, introduced a reassignment process of societal functions:

Since "social worker centre" is stationed in each community neighbourhood of Shenzhen now, we would mobilise these social workers to come to learn the work that we are doing. Once they comprehend this, one day in future if we leave, they can still continue pushing for the same thing. The government at present is also implementing "administration streamline and power delegation", that is to transfer certain functions that the government has assumed over years into civil society organisations to perform, transfer to society to do, society to undertake. The government every year will release a series of funding for us to apply.

(Interview 1, translated)

The "administration streamline and power delegation" is an administrative doctrine and political advocacy recently handed down from the national government. It started an administrative reform in local governments to cut down unnecessary public spending and policy programmes that are no longer relevant. Although this reform is not intended to target on urban sustainability issues, local governments have figured out that it is easier to contract out community and environmental projects than other statutory functions. However, in Shenzhen, the local government is experimenting a new facilitator role, using entities such as the Headquarter Base space, to incubate societal changes that are not necessarily designed by the government anymore. The emergent of grassroots activism thus meet with this recently expanded power vacuum.

To instate the grassroots lobbying power, local NGOs have devised cross-sector initiatives to broaden their societal impacts. For instance, one of the regular practices of the Green Source is to monitor water pollution and to report incidents of illegal industrial discharges. It has been organising voluntary Citizen Watch, in which over 30 local residents living near the polluted rivers were recruited in 2015, to trace down the sources of pollutants along waterways in Shenzhen and to produce written photographic evidence of the pollution status. A "Civil Society Investigation Report on Water Pollution" was thus released to document and communicate the collected evidence. To secure actual changes in industrial malpractices and to give effective pressures of enforcement on local authorities, the organisation hosted a "Shenzhen River Research Cross-sector Report Briefing" in December 2015. These continued efforts have resulted in administrative penalties enforced on three polluting enterprises.

While these polluting enterprises might be easier to deal with than other more powerful actors in urban governance, alliance among grassroots organisations are formed to work on more challenging tasks. In 2013, the then newly founded Green Source and another environmental NGO – China Mangrove Conservation Network – raised concerns that the planned metro extension in Shenzhen would damage a significant area of mangrove forests. They started attempts to intervene such municipal development. However, the municipal office of metro system construction is known as a resource monopoly in Shenzhen's governing system. In addition to making the case by publishing a survey report on local mangrove, the concerned organisations have formed an alliance in monitoring this chain of environmental events to contain the damages on mangrove.

By mobilising voluntary actions and cross-sector supports from local media and nonpartisan political groups, the grassroots activism aims at changing urban management practices, particularly in environmental hazards controls. To be more convincing in persuading changes in local authorities, the grassroots NGO has adopted a unique selling proposition by building a professional staff team trained in ecological and environmental science, who take charge of drafting the mangrove and water pollution reports. It has also incorporated practicing professionals specialised in environmental engineering, green economy and social entrepreneurship as external consultants to the organisational development.

Apart of social innovations, technological ones were also configured in the experimentation of grassroots activism networks. The use of information and communication technolgoy (ICT) has become an inevitable way to disseminate information and to engage with subscribed users, public supporters and volunteers. Moreover, in the Green Source's collaboration with the China Mangrove Conservation Network, a geo-location platform has been launched integrating real-time monitoring of mangrove status with inputs of micro-blogged information. Named as China Mangrove Alert System, such initiative allows voluntary users to perform improvised patrols on endangered mangrove. Users who wish to participate can add hash tags of either "red map" or "black map" to information about mangrove locations and observed risks when publishing their micro-blog posts attached with on-site photos. The text information and photos would be added automatically to the web platform of China mangrove database with geo-located tags crosschecked with information generated from mangrove alert system.[1]

2.2. Civic steering

Whereas Shenzhen as a Special Economic Zone of China – despite being geographically located within Guangdong Province – is not subject to the governing of Guangdong provincial government, other cities in Guangdong are more aligned in regional policy development. Coordination of the provincial government contributes to civic steering of experimental programmes being rolled out in selected urban areas across the province. With explicit attempts of integrating social and technological innovations to attain low carbon development in urban living, a recent plan was designed to construct a Carbon Generalised System of Preferences (Carbon GSP, or *Tanpuhui*) scheme.

Unlike grassroots activism initiatives, the Carbon GSP was designed by regional policy makers with regular inputs from affiliated research agencies. Six cities in Guangdong were called into piloting the scheme: Guangzhou – the provincial capital city, Dongguan, Zhongshan, Huizhou, Shaoguan and Heyuan. A statutory basis for such scheme was formulated and signed off by provincial leadership in July 2015, with a *Guangdong Tanpuhui Pilot Tasks Implementation Plan* covering the overall vision and policy objectives, alongside *Guangdong Tanpuhui Pilot Construction Guidelines* detailing proposed procedures. By policy design, registered citizens in the scheme 'can participate in trading of personalised Carbon Coins on platforms of social media, official website or bespoke mobile application (Box 12.2). These Carbon Coins can be earned by performing designated low carbon behaviours, and can be used as vouchers to redeem commercial services or products.

BOX 12.2 SNAPSHOT OF A CIVIC URBAN LIVING LAB IN CHINA

Name: Carbon GSP (Tanpuhui)
Location: Selected communities or sectors in Guangzhou, Dongguan, Zhongshan, Huizhou, Shaoguan and Heyuan
Geographic boundary: Guangdong Province
Lead organisation: Guangdong Development and Reform Commission
Partners (Types): Provincial-municipal governments, government-organised non-governmental organisations (GONGO), knowledge intermediaries
Running time: 2015–present
Funding sources: Multi-level government budgets
Focus areas: Energy saving, low-carbon travels and communities
Web presence: www.tanph.cn; http://co2bank.cn (beta site)

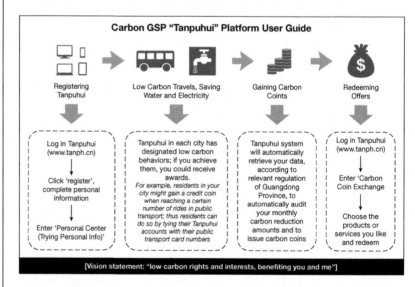

FIGURE 12.2 Carbon GSP "Tanpuhui" platform user guide.

Source: Translated image retrieved from Southern News website.

Local authorities in designated pilot cities are supposed to operationalise this scheme over similar timeframes, using a shared electronic infrastructure devised in the same scheme to provide platform interfaces, to exchange transaction information and to perform carbon data accounting. The leadership intention of designing the scheme was interpreted by one of the core collaborators:

> The motivation came from the governor of Guangdong Province, who raised the idea of incentivising carbon reduction in residents living. He felt that whether we could establish an incentive mechanism to encourage citizens to conduct certain low carbon behaviours; so its core thinking is after I perform a low carbon behaviour, it is similar to one transaction of carbon trading, more or less one CCER,[2] that is after I conduct a low carbon behaviour, I can generate some carbon credits, and then these carbon credits are exchangeable, on which we can also cooperate with businesses, that is to exchange film tickets, or get some discounts, so this is how we want to incentivise carbon reduction in living.
>
> (Interview 3, translated)

The scheme was officially launched for local residents in June 2016 in three pilot cities – Guangzhou, Zhongshan and Dongguan, while the final operation system was still undergoing internal testing. The trial scheme in each pilot city would be operated by its own consortium of actors. For instance, while Guangdong provincial government leads the experimentation in Guangzhou together with other cities, local assistance is coordinated from Guangzhou municipal government. Their technical support was acquired through government procurement from CEPREI, a research laboratory located in Guangzhou that provides verification services. Meanwhile, Guangdong Low Carbon Development Promotion Association – a government-organised NGO (GONGO) established by the provincial government – is responsible for community engagement of the scheme in Guangzhou (Interview 2).

A large part of the consortium formation was administered through procurement. The consortium is thus not strictly speaking a collaborative project partnership, considering that the project implementers are successful bidding suppliers on a principal–agent clientele relationship with the leading organisation, rather than on equal grounds. Additionally, despite open tender procedures, a short notice period was given between the announcement of tender invitation and the closing date for bid submissions (two months for provincial platform building and one month for pilot city project implementation). The procedure was thus preferential to agencies with established programmes involving the government actors, which might limit the potential for bottom-up innovations. Detailed instructions and specific requirements were also laid out in the open tender document regarding the project flows and outputs, so in essence the basic idea and approach were already *designed* by the government before the procurement started.

Nevertheless, the place-based demonstration initiatives for Carbon GSP resonate characteristic of ULL in terms of being geographically embedded, as well as experimenting and learning from sociotechnical innovation in real time. Different low-carbon behaviours are validated for participating pilot cities given the place characteristics. While the amount of household energy saving can be exchanged for the Carbon Coins in most of the participating cities, some behaviours are chosen given the socio-material configurations of different places. In Huizhou, the transactional low-carbon behaviours extend to the cycling distance on bicycles of public rental scheme made available in municipal infrastructure (Interview 3). In Guangzhou, riding public transport can be included because of the wide use of personalised public transport card. Table 12.1 below summarises the geographic boundary of the experimentation within each city, the selected sectors for experimentation and verifiable behaviours.

As indicated by the leadership intention, the Carbon GSP is a mechanism of generating and exchanging incentives. It is thus a social innovation in the sense that it engages with the public in a novel system and structures a new way to drive behavioural change. The general framework and methodology of Carbon GSP,

TABLE 12.1 Carbon GSP place-based experiments

Geographic boundary	Sectors	Low-carbon behaviours	Involved users
Guangzhou	Building, transport	Saving residential energy/ water; buying green building property; reducing private car use and taking public transport	General public
Dongguan: Civil Servant Community	Housing, hospitality	Purchasing selected energy-efficient products; sorting and recycling wastes	Selected residents
Zhongshan Heyuan: country park, industrial park, low-carbon hotels	Commerce Tourism, manufacture	(Not specified) Visiting the country park with energy retrofit and carbon sequestration forests; purchasing selected energy-efficient products; staying in low-carbon hotels	SMEs Tourists
Huizhou	Community, transport	Cycling with hired municipal bikes	Local residents
Shaoguan	Forestry	Preserving forests from urban development (eco-compensation)	Local residents

Source: Tanpuhui Open Tender Document (November 2015)

together with who should be the actors to deliver and operate the system (as "duties"), have been meticulously stipulated by provincial-level policy makers. Concrete steps were proposed in the initial policy design in 2015 (*Guangdong Tanpuhui Pilot Construction Guidelines*), set out as:

1 selecting pilot communities;
2 choosing low carbon behaviours;
3 quantifying carbon reductions in residents' low carbon behaviours;
4 retrieving relevant data of low carbon behaviours;
5 confirming incentive methods for low carbon behaviours;
6 issuing Carbon Coins and exchanging Carbon Coins;
7 hosting promotional activities;
8 assembling total volume of carbon reduction, and joining it to the carbon emission trading system (ETS).

The social innovation is embedded in mainly step (5) and (6) above, devising an entity of Carbon Coins to create incentives. The experimentation of such social innovation is thus concerned with how to activate and sustain effective flows of the Carbon Coins.

To operationalise this design, the Carbon GSP not only incorporates ICT engagement techniques, but also requires technically a method to aggregate behavioural data acquired from multiple sources and entities via the electronic infrastructure and to share these data across the province. The ICT techniques were mainly used in configuring civic interfaces of the new scheme to generate incentives and facilitate real-time interactions.

To retrieve relevant behavioural data as suggested by step (4) above, the Carbon GSP should be able to acquire data from utility and public service provides that are often formatted and/or stored in different ways. It should also be capable of automatically integrating these different data from their sources before importing them into the Carbon GSP accounting system. The data integration process is thus also a process of negotiating and engaging with various stakeholders of the city to gain their consents on information access. These stakeholders include water suppliers, district-level power distributers, gas companies, property management companies (who can provide monthly records of vehicle access to infer on the use of private cars), bus companies, public transport card issuing company, traffic operation companies and/or traffic control center. Some other individual consumption or travel behaviours (i.e. visiting country parks, riding eco-cars, staying in low-carbon hotels or purchasing energy efficiency products) are to be collected by scanning Quick Response (QR) Codes to set off real-time data transfer.

Through this experiment, further policy questions pondering are how to connect the accounting and transaction system of Carbon GSP with the established provincial carbon emission trading system (ETS), while the latter is also undergoing experimentation. Learning from the Carbon GSP regarding self-governing responses at individual or community levels might be translated into the context of governing through mandated enabling in the ETS that attempts to shape industrial activities. Referencing such learning, provincial-level policy makers seek to reflect on how to expand the scope of the current ETS.

2.3. Strategic controls

Strategic experimentations involving more national government inputs have displayed similar characteristics of the civic initiatives discussed in the last section, particularly regarding policy mandates and administrative procedures such as the use of procurement mechanisms. However, they appear to focus more on technological development and enabling top-down distant governing controls, while less about community empowerment or devising social innovation.

Place embeddedness thus turns out to be less relevant, when the location of experimentation can be replaced and actors are easily migrated within institutionalised national networks of organisations. The smart city programme in Nansha – a nationally endorsed "state-level new urban district" and "free-trade experimentation zone" located in Guangzhou – demonstrates some of the distinctions of a strategic experiment.

The vision and discourse of Nansha's smart city programme have been weaved into various policy narratives such as governmental guidelines or nongovernmental evaluation frameworks of smart cities, while its design and practices are mandated by policies of multilevel authorities. The statutory basis that the experimental program operates on is thus tighter than either civic or grassroots initiatives. It controls not only the program expectations, but also the eligibility of leading organisation.

The smart city programme in Nansha is being led by the Guangdong Innovation Alliance of Smart City Industrial Technology, a GONGO formed with a provincial policy mandate enacted in 2009. The policy called for governmental and nongovernmental actors to initiate an "Industrial-Academic-Research Alliance" type of organisations. The initiated "alliance" would accommodate partnership activities between industries, academic institutes and local government. The Guangdong Innovation Alliance of Smart City Industrial Technology was thus formed by Chinese Academy of Sciences' software research institute that is based in Nansha, together with other state-led and corporate-led research institutes. To continue leading smart city programme in Nansha, the "alliance" organisation is subject to annual review of the scientific department in either municipal or provincial government. Governmental regulators, who conducted on-site examinations in the "alliance" organisation, hold the administrative power to decide on whether the "alliance" should be terminated, be remedied or receive core government funding.

Under on-going performance monitoring, the "alliance" organisation in Nansha had coordinated a range of initiatives within the umbrella programme of Smart Nansha. The collective of research institutes launches experimentations on Data Hub architectural design, cloud-enabled district administration, Internet of Things (IoT) infrastructures as well as Smart Communities constructs. These initiatives are experimented with a district-level control room on urban management, multi-

BOX 12.3 SNAPSHOT OF A STRATEGIC URBAN
LIVING LAB IN CHINA

Name: Smart Nansha

Location: Nansha New District

Geographic boundary: Guangzhou Municipality

Lead organisation: Guangdong Innovation Alliance of Smart City Industrial Technology

Partners (Types): National government, state-led knowledge intermediaries, technology innovation firms

Running time: 2009–present

Funding sources: Multi-level government budgets; research grants; corporate investments

Focus areas: Smart communities supported by Internet of Thing infrastructure; Data Hub supported urban governing; digital platform-enabled administration

FIGURE 12.3 State-led research facilities and showcase of smart lamp-post in Nansha (field visit, May 2016)

purpose sensors installed on street lampposts standing in the newly urbanised part of Nansha, as well as other technical supports either acquired from procurement or provided by "alliance" partners.

The technological attention is accompanied with relatively few considerations of community engagement. The Smart Communities in essence refer to constructing a digital platform gathering community management data, alongside some physical user terminals to be installed in local communities. Actual inputs and participation of citizens are marginal in Nansha's smart city construction process that positions local residents as end users of developed technological innovations rather than potential co-creators. The umbrella programme of Smart Nansha has been formulated based on top-down system design carried out often by local research branches of national academies. It thus tends to stress on strategic planning rather than addressing social needs of local communities.

Social, institutional and technological innovations are considered discrete responsibilities of different actors. Depicted by the initiator of the government-steered smart city alliance, passive change making was adopted by this GONGO:

> For institutional innovations, we can only make some suggestions; institutional innovations eventually still depend on the government, we are only positioned as a think tank, giving some advices, but eventually it's quite often that the fundamental institutional innovation is still held in the hands of the government. For it (the government) may face certain problems, facing problems then after all such institutional mechanism – the filtration of this organisational structure is still required. Technological innovations would then be undertaken by technical work unit (*danwei*), such as research institutes, or enterprises to work on technological innovations, so then the government is not responsible for technological innovations. The government is mainly arranging fabrics of the process, organising a structure (in terms of) how it benefits the citizens more.
>
> (Interview 4, translated)

Nevertheless, strategic experimentation might have strong impacts in processing change. Channels of learning have been configured in both formal and informal national networks of knowledge organisations and local authorities of the proclaimed smart cities, with smart urbanism becoming the mainstreamed way of local development in China. Meanwhile clear plan for scaling up Nansha's experiment had been written in various policy documents before the local vision was formulated. The Smart Nansha trial run would be expanded into Smart Guangzhou, particularly designed in a municipal implementation plan published in 2012.

3. Re-designing mechanisms of managing cities

Grassroots initiatives are often geographically entrenched with local resources, tacit knowledge of community leaders and existing socio-material configurations in both natural and community environments, so they are hardly transplantable to another location without substantial adjustment to its design. They generate mostly *self-governing* (Bulkeley and Kern, 2006) responses to sustainability challenges. Yet, several new ways of engaging people and communities are seen in mobilising resources, implementing the experimentation, and disseminating the results. These practices also exhibit explicit attempts of making bottom-up changes to urban management practices.

Civic initiatives, as proposed by actors from regional and local governments, reflect upon place characteristics in their design. They are not, however, restrained in their own localities of origins, and they are configured in transferable forms for the sake of policy learning. Policy mandates are still constituted as a necessary governing tool to form civic initiatives and to ensure compliance from core local stakeholders. *Governing by authority* is thus evident. The Carbon GSP scheme in particular does demonstrate an organic mixture of social and technical innovations, which integrates the governing by authority with supplementary devices of *governing through enabling*.

Strategic initiatives tend to be formulated with multilevel mandates, while place embeddedness is less of a determinant than in the civic and grassroots cases. Technological innovations are being experimented in purposefully selected urban sites, not to address social needs but to test out specifications of these innovations and how they can "land on the ground" (Interview 4, translated). *Governing by authority* is effectively combined with *governing with provisions*, where citizen involvement is only relevant in disseminating these provisions. Social innovation exists in devising new ways of urban management made possible through technological innovations, such as cloud-enabled administration and IoT infrastructures, but not in terms of engaging society.

Social–technical splits are therefore still discernible in the design and practices of urban experimentation. The array of grassroots, civic and strategic initiatives display a spectrum of this socio-technical divide. Figure 12.4 illustrates how such a divide is linked with the varied intentionality of discrete networks of actors. Institutional entities at the national level tend to invest in and align around strategic initiatives; while the municipal and provincial (mid-range) governments are often leading civic initiatives together with universities based in the same localities, the mid-range governmental actors are often the agents of change implicitly designated by the national government. As to the grassroots initiatives that often fall away from the limelight, local actors of change from a wide range of sectors (not limited to the government agencies) are driving their designs and practices. Such variation in the design of strategic-civic-grassroots ULL in the observed part of China is also brought in due to a split of interests in advancing social and technical innovations: with the strategic ULL inclining to purely technological innovations that require blocs of capital investments, civic and grassroots ULL tend to concentrate on local community interests that are leading to the improvisation of social innovations rather than a confined approach of technological inventions.

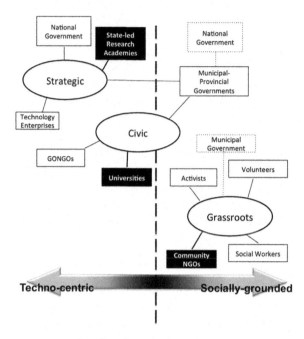

FIGURE 12.4 Social–technical split and networking divides

The multiplicities of co-existing experimentation forms of grassroots, civic and strategic natures are starting up dynamic processes of urban transitions with both bottom-up and top-down flows of intentionality, resources, mandates and in certain sense powers. As in the relative process of change, the collective outcomes of such multiplicities indicate a stage of re-designing urbanity, rather than radical transformation as such. A practical way of re-designing is evaluating current practices to enable learning for effective changes. Intertwined with the funding models in China, evaluation becomes another differentiating characteristic across the discussed types of initiatives.

The practices of evaluation are probably institutionalised in grassroots initiatives as matters of survival. Distinct from GONGOs that securely receives regular governmental resource supports as trusted agents to deliver income-generating public services, independent NGOs operate on seeking alternative financing. Moreover, the financing model of community development trusts (commonly seen in urban communities in OECD countries, e.g. see Gilroy, 1996, Sahd, 2004) by regenerating community-held assets has not been widely instituted in China. Consequently, regular evaluations have become a norm. The intertwined relationship between evaluation and funding was revealed as:

> [1a] We always have evaluations; there are always evaluation agencies; we need to conduct evaluations on ourselves as well for the application of some

funding . . . for all of our funding is sourced externally. [1b] Right we don't hold any assets ourselves.

(Interview 1, translated)

When further explained, evaluation practices that had been carried out by third-party assessment agencies often followed the evaluation framework formulated by the funders when they set out the funding programmes. Social impacts have become a frequently assessed component, measured by various indicators, such as whether the projects had received any awards, or by interviewing targeted beneficiaries about their thoughts regarding the projects. Evaluation reports were compiled and submitted to the funders, who thus could decide whether the funded organisations would receive the next phases of funding.

A primary source of funding for the Green Source had been non-polluting and locally-based enterprises that had been keen supporters of the citizens-led pollution monitoring initiatives. Other alternatives were nongovernmental charity foundations, which however had cut down resources due to financial pressures on themselves. District-level governments were another resort, which gave out more community funding than the municipal government. The supports of district governments and local enterprises are again making grassroots initiatives more geographically entrenched.

By contrast, strategic and civic initiatives are secured financially and the evaluation practices turn out to be more informal. There is no self-evaluation practice instituted or documented within the leading organisation of Smart Nansha, apart from the occasional visits paid by governmental regulators to examine the organisational performance and project progress. Self-evaluation practice is deemed "unnecessary" and "without particular purpose" by one of the leading individuals of Smart Nansha, while the occasional visits were interpreted as governmental "care" rather than challenges to productivity (Interview 4, translated). This sense of security is probably derived from the relatively steady funding flows, alongside explicit political commitment made by multilevel authorities that are drawing rising interests from industries and innovation firms.

Evaluation is conducted, however, on project partners that the leading organisations of civic and strategic initiatives have procured. Such evaluation is an assurance mechanism institutionalised in governmental procurement procedures. With the lack of evaluation framework set out in the midst or after the experimentation, performance assurance relies on the procurement evaluation imposed at the start. Moreover, project delivery cannot be guaranteed by the way contractual payments are made. Failures of delivery are only subject to a penalty of 5 per cent (the sum of last payment) of total contractual value, when the rest of the total payment would have already been issued at the project launch.

Embryonic characteristics that differentiate the varied forms of experimentation in this particular regime context are identified in the above text, regarding their geographical embeddedness, governing responses and intensities of social innovations. These various characteristics are listed in Table 12.2. They should be

TABLE 12.2 Characteristics of ULL typology in China

	Grassroots	*Civic*	*Strategic*
Funding patterns	A mix of government subsidies, project service incomes, membership fees and donations	Government earmarked budget (procurement); commercial sponsorship	Government earmarked budget (procurement); corporate resource supports
Leadership and ownership	Community-owned initiatives led by independent NGOs	Owned and led by local authorities and GONGOs	State-owned experiment, co-led by GONGOs, regional government and SOEs
Geographical embededness	High	Medium	Low
Governing responses	Self-governing	Governing by authority and through enabling	Governing by authority and with provisions
Technical innovations	Yes: For dissemination and engagement	Yes: For operation, engagement and dissemination	Yes: Centre piece of operation and for managing city
Social innovations	Multiple	Core design	Few
Use of ICT	Yes	Yes	Yes
Evaluation mechanism	Yes: For funding continuity	Yes: For procured project delivery; policy learning	Yes: For leadership eligibility; procured project delivery
Changes in urban landscape	Material change in urban environment	New infrastructural design	New logics of urban management
Policy implications	Bottom-up change	Designed for policy learning	Strong due to core governmental involvement
Translation/ scaling up	No plan on scaling up	Clear policy plan on translation	Clear policy plan on scaling up
Institutionalisation/ main streaming	The model of the Green Source is still a rather unique case currently	It depends on pilot project outcomes. Similar initiative has been proposed in Hubei province	Smart city initiatives have already become a mainstream local development approach in China

considered as *inductively* structuring an emerging typology of ULL in China, rather than a deductive evidence building one.

Table 12.2 summarises initial findings from comparing the three types of ULL. The main distinction in funding patterns exists in grassroots initiatives, whereas civic and strategic ULL share similar way to funding. Grassroots ULL, given its low political significance attached, are still searching for the appropriate funding forms and partners. Identifying the whereabouts of leadership and ownership might be a generic shortcut to differentiating the typology of ULL. Place factors define the grassroots ULL, but their importance reduces as level of tactics increase. Forms of governing responses vary across types of ULL, but this might not be generically applied to all cases and thus governing responses should be examined in case-by-case manner. ICT and other forms of socio-technical innovations are widely used in all range of ULL, but they are used for different purposes across types of ULL; the exploration of both the forms and designed goals of fitting in socio-technical innovations in ULL is thus necessary. Evaluation has been necessary for all types of ULL discussed, but again such evaluation is for different purposes across types of ULL. Landscape change varied by scales, from partial display of urban materials, to infrastructures and to management models. Policy implications of ULL varied given the distance to governing administration, but the administrative barriers are not irremovable. The intention to translate the ULL into other contexts, or to scale it up to upper levels of governance, is not discernible in grassroots initiatives but it is often essential in civic and strategic ones. The outcomes of whether ULL as innovations and unique cases can be institutionalised into political structures, or mainstreamed into broader culture, are relatively difficult to assess systematically with limited resources deployed in this chapter, which should be a subject for further research.

4. Conclusion

The comparative case analysis in this chapter has discussed several common strands of designs and practices in the multifaceted urban experimentation in China, a nation-state that has been defined as a "semi-authoritarian regime" (Ho, 2007; He and Thøgersen, 2010). So whether and how experimentation processes collectively shape the democratic space for urbanity remains a question. The insights draw from case studies of ULL might serve as conceptual probe for further research on the subject matter of experimentation and (urban) state building.

The common and wide use of ICT arrangements across different types of ULL initiatives represent a compelled change in power domination. Political power has become more dissolved into individual hands, without any radical change in the political regime. Voices from bottom up in a polity are leaving stronger footprints. Community-owned grassroots initiatives are inducing new ways of generating self-governing responses to sustainability challenges. They also denote emergent forms of power, met with a public administrative paradigm of "administration stream-line and power delegation" recently proposed by the current national leadership in China.

The active use of ICT or other forms of socio-technical innovations in ULL might as well be an uptake of means to change, by agents of change, be they the governments, their oppositions or organised groups of the general public. As simply a means to change, the innovations deployed are added with socio-political meanings derived (or deviated) from their designed goals. Despite the design framework, the innovations as means to change are casting result in actual outcomes in society that derail from the initially designed goals. The evaluation of the actual outcomes of ULL as compared to their original design is thus necessary, on which we should take the broader institutional context into account.

Among the different types of ULL initiatives discussed in this chapter, the grassroots ones represent a new age of, and are configured with evolved approach to, driving changes in urban transitions in China. Grassroots activism is transforming the established entities and institution of urban governing, with new ways of (1) engaging both citizens and local authorities, of (2) lobbying for new urban management practices, and of (3) setting a political agenda to secure consistent funding that share similar visions. Yet, such transformation is geographically embedded, and is thus confined within a certain scope of geographic boundary. Harnessing local government support is still essential for organisations leading bottom-up initiatives, in terms of generating alternative project incomes, securing office space and making effective policy changes.

Manifested in the different types of ULL initiatives, an element of change making is formulated as a main goal, either about driving individual behavioural changes or about transforming urban management practices. It is pertinent to note that contrasting the proactive approach adopted by independent NGOs, rather passive approach to change making is adopted by GONGOs that are often leading or co-managing civic and strategic experiments. After all, despite the observed socio-technical divide (Figure 12.4), the multiple directions of changes driven by the emergent of diversified urban experiments to some extents create a more democratic space for transformation.

ULL in general carry a built-in vision of urban sustainability transitions, which can be generalised across different socio-political regime contexts. In a socialists' state context, it is fundamentally seeking "paths in Utopia" (Buber, 1949, reprinted in 1996), whichever extents market components have been incorporated in the practices to finding it. In the development and rapid urbanisation process of south China cities, citizens have been stepping up to rebalance environmental interests of the society against pure economic growth. Meanwhile, these initiatives have been approached by various entities of state or market apparatus to advance social innovations with technological progress.

Notes

1 The interactive mangrove map can be accessed at www.china-mangrove.org/common/map/big_map
2 CCER is abbreviation for Chinese Certified Emission Reductions.

References

Buber, M. (1996). *Paths in Utopia: Martin Buber 1878–1965*. Syracuse, NY: Syracuse University Press

Bulkeley, H., Breitfuss, M., Coenen, L., Frantzeskaki, N., Fuenfschilling, L., Grillitsch, M., . . . Voytenko, Y. (2015). Theoretical framework: Working paper on urban living labs and urban sustainability transitions. *GUST Project Publications*. Retrieved from www.urbanlivinglabs.net/p/publications.html (accessed 1 October 2016).

Bulkeley, H. and Kern, K. (2006). 'Local government and the governing of climate change in Germany and the UK'. *Urban Studies, 43*(12), 2237–2259.

Evans, J. and Karvonen, A. (2010). 'Living laboratories for sustainability: Exploring the politics and epistemology of urban transition'. In Bulkeley, H., Castán Broto, V., Hodson, M. and Marvin, S. (Eds.), *Cities and Low Carbon Transitions*. London: Routledge.

Gilroy, R. (1996). Building routes to power lessons from Cruddas Park. *Local Economy, 11*(3), 248–258.

Guangdong Development and Reform Commission and Guangdong Provincial Government Procurement Centre. (2015). Government Procurement Agency Agreement [public document].

Guangdong Provincial Government Procurement Centre. (2015). Tanpuhui open tender documents [public document].

He, B. and Thøgersen, S. (2010). 'Giving the people a voice? Experiments with consultative authoritarian institutions in China'. *Journal of Contemporary China, 19*(66), 675–692.

Healey, P. (1998). 'Building institutional capacity through collaborative approaches to urban planning'. *Environment and Planning A, 30*(9), 1531–1546.

Ho, P. (2007).' Embedded activism and political change in a semiauthoritarian context'. *China Information, 21*(2), 187–209.

Hodson, M. and Marvin, S. (2007). 'Understanding the role of the national exemplar in constructing "strategic glurbanization"'. *International Journal of Urban and Regional Research, 31*(2), 303–325.

Mai, Q. and Francesch-Huidobro, M. (2014). *Climate Change Governance in Chinese Cities*. London: Routledge.

Sahd, B. (2004). 'Community development corporations and social capital'. In R. M. Silverman (Ed.), *Community based organization: The intersection of social capital and local context in contemporary urban society* (pp. 85–122). Detroit, MI: Wayne State University Press.

Silver, J. and Marvin, S. (2016). 'Emerging styles of urban experimentation'. In J. Evans, A. Karvonen, R. Raven (Eds.), *The experimental city*. London: Routledge.

Voytenko, Y., McCormick, K., Evans, J. and Schliwa, G. (2016). 'Urban living labs for sustainability and low carbon cities in Europe: Towards a research agenda'. *Journal of Cleaner Production, 123*, 45–54.

List of cited interviews

Interview 1: [a] Chairman, [b] Fund Raising and Communication Officer, Shenzhen Green Source Environmental Protection Volunteers Association, November 2015, Shenzhen, China.

Interview 2: Secretary General, Guangdong Low Carbon Development Promotion Association, May 2016, Guangzhou, China.

Interview 3: Deputy Director, SYSU Guangdong Research Centre for Climate Change, May 2016, Guangzhou, China.
Interview 4: Director, Guangdong Innovation Alliance of Smart City Industrial Technology, May 2016, Nansha, Guangzhou, China.

Acknowledgement

The fieldtrip in China conducted in May 2016 was partly funded by Regional Studies Association (RSA); the author acknowledges the support of Prof Alex Lo at the University of Hong Kong.

13

THE IMPORTANCE OF PLACE FOR URBAN TRANSITION EXPERIMENTS

Understanding the embeddedness of urban living labs

Frank van Steenbergen and Niki Frantzeskaki

1. Introduction

The imperative of experimentation as a means of planning and policy testing in cities has been showcased over the past years especially for unlocking governance innovations for sustainability. Experimentation as a form of governance for sustainability transitions has been advocated from transition scholars (Frantzeskaki, Loorbach and Meadowcroft, 2012; Frantzeskaki, Wittmayer and Loorbach, 2014; Smith and Raven, 2012; Pereira, Karpouzoglou, Doshi and Frantzeskaki, 2015; Wolfram and Frantzeskaki, 2016) and from geography scholars (Castan-Broto and Bulkeley, 2014), showing that intervening to governance dynamics in cities with new lessons and new forms of innovation has the potential to reroute urban development to more sustainable outcomes. Even though experimentation can be realised through various formats and across different scales, we chose to examine the setting of urban living labs (ULL) as a format of urban experimentation. ULL focus on stimulating certain processes via which governance experimentation is designed, tested and implemented (Voytenko, McCormick, Evans and Schliwa, 2015). ULL are aimed at co-creation and empowerment of multiple stakeholders in co-shaping an experimental approach (Edwards-Schachter, Matti and Alcántara, 2012; Lehmann, Frangioni and Dubé, 2015) and being open and participatory (Franz, 2015). The experimentation in these labs is often placed in a geographical area in the sense that they represent ecosystems of open "urban" innovation, and are situated in a real urban context where the process in focus is taking place. This may be a region, an agglomeration, a city, a district or neighborhood or a building. ULL are also tied to a certain local context that surrounds the "lab" in question and determines its (symbolic) meaning.

While early research, examined ULL as test beds for urban technological innovation, the focus has shifted recently on other forms of innovation including social innovation (Franz, 2015; Edwards-Schachter et al., 2012). A socially centered approach to understand and examine ULL as Franz (2015, p. 58) addresses it includes an understanding of the processes of "translation" and "contextualisation" also considering that ULL aim at making innovation inclusive and function as spaces of encounter. Taking place as a factor for translation and contextualisation can deepen the conceptualisation, examination and design of ULL, as socio-spatial interventions alongside with as governance innovation instruments. Against this understanding we propose that examining the impact of ULL in urban sustainability transitions requires an understanding of its embeddedness in socio-spatial context.

The notion of embeddedness or "sense of place" is rather under-examined in sustainability transitions literature. Although, there has been an increasing attention to geographical and spatial dimensions within transitions literature (a foci on cities and nations) in recent years (Binz, Truffer and Coenen, 2014, Coenen, Benneworth and Truffer,, 2012), there is a hiatus that is important to address in regard to the contextualisation of experimenting. Thus, it remains unclear in what way place-embeddedness influences the impact of urban experimentation and it follows that unexplainable research outcomes are often related to the little satisfying black box of context-dependency. And it is exactly this hiatus in the literature that we aim to address in this chapter. By drawing on literature on insights from different disciplines ((urban) sociology, geography and anthropology) in regard to a "sense of place" (Gieryn, 2000; Gieseking et al., 2014), we address the following research question: *in what way does place-embeddedness influence the design, practices and processes of ULL as experimentation settings for urban sustainability transitions?*

In this question, we perceive the geographical place not as a given, but engage with it from a relational perspective, that is as a fluid category constantly under debate. In order to fully grasp how ULL are tied to this fluid geographical setting, we try to address in how far a ULL is being reframed as to fit a current particular context and in how far it is being closed down and imposed on a specific context.

2. Understanding embeddedness: Urban living labs as transition sites

To understand embeddedness as a process of integration of an experiment in a geographical context, writings on sense of place in cities are our conceptual basis. The notion of embeddedness or "sense of place" is rather under-examined in ULL literature. Although, there has been an increasing attention to geographical and spatial dimensions within, for example transitions literature (a foci on cities and nations) in recent years (Binz et al., 2014; Coenen et al., 2012), there is a hiatus that is important to address in regard to the contextualisation of experimenting in ULL.

Sense of place is "the collection of meanings, beliefs, symbols, values and feelings that individuals and groups associate with a particular locality" (Williams and Steward, 1998). Restorative topophilia "represents an opportunity for positive dependence that underpins the emergence of virtuous cycles" in urban social-ecological systems, relating to strong sense of place (Tidball and Stedman, 2012, p. 297). It relies on creating or strengthening new community relations with place and as result of these relations and new meanings of place, place characteristics and urban place capacities are regenerated. As thus, topophilia is "constructed" (Tidball and Stedman, 2012). So, attachment to place and (re)creating sense of place can be the result of learning and social constructed meanings facilitated in urban experimentation settings like ULL, and ultimately, result in and lead to place transformations. This extends the understanding of Stedman and Ingalls who position topophilia "as a powerful base for individual and collective action that repair and/or enhance valued attributes of place" (Tidball and Stedman, 2012, p. 297). We conceptualise that sense of place can be an outcome of experimentation that leads to urban sustainability transitions, when narratives and new relations within sense of place include elements of a theory of place transformation.

From the writings on sense of place we distill that there are three overarching phenomena that relate to a strong sense of place as a medium to (place) transformations.

2.1. Changing and contested narratives of place

A narrative of place, which connects historical routes with place-change visions. Multiple narratives of place can co-exist showing the contested understandings of place from different members of a community. As Chappin and Knapp (2015, p. 39) argue, "sense of place is often contested and not a simple panacea for stewardship, as sometimes assumed by environmental advocated". A narrative or narratives of place can illustrate the multiple understandings and shed light to "the complexity of sense of place" considering that it can be both instrumental and detrimental for change. In our conceptualisation of sense of place as catalytic for urban transformations, we seek to discover narratives of place that encapsulate change visions or elements of theories of change in them.

2.2. Symbolic understandings of place

A symbolic understanding or meaning of place that captures the sentiment of the community about the place of interest often shows the "sense of belonging" and encapsulates the different local experiences (Wilbanks, 2015, p. 76; Tomaney, 2014). The meaning of the human experience and the emotions and thoughts accompanying it provide importance of place (Stedman, 2003). A symbolic meaning does not necessarily relates with a spiritual meaning, but rather shows that a plurality of symbolic meanings can coexist "leading to different attitudes, intentions and actions (. . .) despite shared appreciation for the same biophysical features" (Chappin and Knappe, 2015, p. 41). From the focus of sustainability transitions, we align with the sociocultural understanding of place meaning "that shifts (. . .) from inherent and instrumental forms of meaning towards a new meaning as socially or symbolically constructed within the cultural, historical, and geographical contexts of day-to-day life" (Williams, 2014, p. 76). As such, urban experimentation processes contribute to learning and shifting meanings by contributing to socially constructing them.

2.3. Place-related relations and networks

New types of relations between people and place, for example restorative, attached, detached are created that reflect also personal meanings as well as "the power relationships that shape historical and current interactions with places" (Chappin and Knapp, 2015). These relations further relate with the materiality of place on how resources in a specific place are being used, accessed and divided (Williams, 2014). As such, we propose that sustainable relations between people and place can provide a normative orientation of sense of place as a transformative medium. Urban experiments contribute to new relations of people and place in the sense that "made, unmade and remade in relation to human projects" these relations (Entrikin and Tepple, 2006).

Based on these overarching phenomena, we conceptualise ULL as "transition sites". However, one of the basic conditions for a ULL is then, that it is embedded in the (contested) narratives, symbolic understandings and relations and networks in a certain place.

2.4. Translation techniques of ULL

In order to assess this, we discerned the following five translating techniques related to the changing narratives (1 and 2), symbolism (3 and 4) and networks (5) which ULL can use in order to embed their design, practice and processes:

1   Local entanglement: in what way is the lab entangled in local historical roots and discourses? And how does it relate itself to the dominant narratives?
2   Contextualisation of the lab: how does the ULL contextualise their methods and practices to the characteristics of a place? And how does the experimenting take place on an everyday basis?
3   Framing its distinctiveness: how does the ULL frame their distinctiveness and/or innovative character? In what way does it take the different symbolic meanings of the place into account?
4   Setting the scene: how does the ULL set the scene for their interventions? Why is the lab placed in a certain context? And how is it perceived by local actors?
5   Participation and collaboration: who is participating in this experimenting and how? What kind of collaborations in and with the ULL are taking place? And

how does this participation and type of collaborations influence the design, practice and processes of the ULL?

With taking these techniques into account, we hope that we can discern the notion of embeddedness of ULL employing the conceptualisation of sense of place as an analytical device to understand embeddedness of experimentation that takes place in the ULL and the implications this raises for current and future research and practices of ULL.

3. Approach: Longitudinal place-based research

In order to make a first attempt in doing so, we examine the case of Veerkracht Carnisse, an urban regeneration intervention that focuses on empowering local communities and making the area more sustainable and resilient in the period of 2011–2015. It is interesting to look at this case since Veerkracht can be considered as an ULL that is based in the district Carnisse in the city of Rotterdam, which is often framed as a "deprived urban area" and where a multitude of problems are often highlighted (e.g. poverty, low level of cohesion, poor housing, migration waves, etc.). Due to this specific context, we aim to address how far the Veerkracht Lab contextualises its design, processes and practices to characteristics of the district in order to be innovative and tap into the local dynamics.

The research used for this chapter is part of longitudinal research activities of DRIFT at the Erasmus University in Rotterdam. Based on the notion and approach of action research, several researchers have taken part in activities of the living laboratories. For instance, they have applied the transition management methodology in Carnisse, which resulted in the Community Arena (supported by the FP7 project InContext, 2011–2014). Besides being actively involved in the field and contributing to several experiments, researchers were primarily responsible for the monitoring of the living lab.[1] In this research roughly five different kinds of actors can be discerned, namely: members of the Veerkracht-consortium on a strategic, tactical and operational levels; participants in the activities of the lab such as volunteers, children, families, teachers; inhabitants of Carnisse (or the broader district of Charlois); neighborhood professionals like welfare workers, civil servants, policy makers, youth coaches, social workers and other participants like urban experts and professionals and local actors in other districts. This distinction in actors is relevant in order to understand which kind of perspective respondents have with regard to Carnisse and the Veerkracht Lab.

The data for this chapter however, is primarily derived from research in the period of January – September 2015, gathered during the research activities within a JPI Urban Europe funded project, called GUST.[2] The following empirical data are used: 34 interviews with different actors, four monitoring sessions with lab practitioners, one participatory workshop on impact of Veerkracht Lab, eight monthly progress meetings and numerous amount of (participatory) observations and numerous field visits.[3]

4. The case of Veerkracht Carnisse, Rotterdam, The Netherlands

The case of Veerkracht Carnisse, or Resilient Carnisse (*in English*), is an ULL that focuses on empowering local communities and making the area more sustainable and resilient in the period of 2011–2015. It is implemented in the neighborhood Carnisse of the harbour city Rotterdam, the Netherlands. Carnisse has around 11,000 inhabitants and is part of the larger district of Charlois (with about 65,000 inhabitants). "Veerkracht" stands for resilience and is a consortium of four partners: Rotterdam Vakmanstad, Creatief Beheer, Bureau Frontlijn and DRIFT. Each of these partners had their own particular focus, activities and methods more insightful. Veerkracht was quite an open and diverse lab. With the very different activities and foci, it is hard to draw a clear boundary around the lab when looking at the breadth of the practices initiated and facilitated by the lab partners. As stated, in geographical terms it can be said that the primary focus was the neighborhood of Carnisse and secondary focus the larger district of Charlois.

The goals of Veerkracht Carnisse were to strengthen the resilience of the neighborhood, to empower children, families and communities and to gain insights in necessary reforms of the current district planning of the city of Rotterdam. These goals were rather broad and ambitious. Different experimenting methods were employed in the Lab to allow a mutual learning and sharing of information to achieve the goals.

The target groups of the combined actions include:

- Primarily: children (4–12 years), their direct families, schools (board, teachers and parents) and residents or volunteers actively involved in community life
- Secondary: the different (institutional) networks in Carnisse (and Charlois) consisting of professionals, civil servants, social workers, entrepreneurs, etc.

Veerkracht had an official running time of five years. It started with a period of concept development and scoping in 2009, and officially kicked off in September 2011 and ended in August 2015. From August 2015 until today, there is a continuous collaboration with the Lab partners in searching for new ways to sustain strengthening the resilience of the neighborhood. The lab activities focused on: poverty reduction, the upbringing of children, democratic reform, support of biodiversity, etc. The main element of experimentation is the application of the different engagement and participatory methodologies in the local community.

The partners and participants involved in the Veerkracht program did not explicitly call themselves an "urban living laboratory" or a "lab". They addressed Veerkracht as "an experimental program" and within its spatial and administrative boundaries tested new methods and practices. They discerned four different "fields of interaction": home, school, outdoors and neighborhood. Each of these fields of interaction consisted of individuals, networks and institutions on which the activities were focused. By working together in Veerkracht the aim was to increase the interaction between the different target groups via the engagement in these four fields of interaction and in this way, creating a more integral way of working for transformation at the scale of an urban neighborhood. The assumption behind Veerkracht was that this interaction took place via collaboration in practice and created an added value in a financial, social and ecological sense (when compared to traditional ways of neighborhood development).

4.1. Local entanglement

Deprived urban neighborhoods in the Netherlands have been increasingly targeted for "revitalisation". This is especially the case for the city of Rotterdam, where the "neighborhood discourse" has been increasingly popular in the last two decades. The southern neighbourhood Carnisse with its close to 11,000 inhabitants is considered to be one of the forty most "disadvantaged neighbourhoods" or "special interest neighbourhoods" in the Netherlands (Ministry of Housing, Spatial Planning and Environment, 2007).[4] While governmental interventions in Rotterdam have mainly focused on physical renewal, with the slinking budgets in the aftermath of the economic crisis in 2008, this has not been an option anymore. Slowly, the national neighbourhood approach changed and more emphasis was put on the role of inhabitants and citizens, who should become more active in addressing and solving problems in their living environment (Visitatiecommissie Wijkenaanpak, 2011). This shifting understanding is part of a broader discourse on the changing roles of citizens and governments in what came to be known as "Big Society" in the United Kingdom (Kisby, 2010; Ransome, 2011) and "participation society" in the Netherlands (Putters, 2014, Tonkens, 2014).

Due to several waves of popularity in neighbourhood policy discourses (Uitermark, 2003), there is a long tradition of interventions aimed at addressing living conditions targeting safety, clean streets, social cohesion, cultural diversity, poverty-reduction and more. Therefore, residents in neighbourhoods like Carnisse were rather familiar with these interventions and have increasingly grown weary of them (Van Steenbergen and Wittmayer, 2012). A cultivated mistrust towards outsiders (like entrepreneurs, professionals, civil servants, researchers, etc.) was combined with a critique of top-down approaches that stigmatise the neighbour-hood and its residents that have become part of the problem that these approaches actually want to address. In the last decades several revitalisation strategies have been carried out in deprived areas. Most of these strategies can be typified as being top-down policies with a strong emphasis on physical and economic pillars. For example, the strategic demolition and renewal of housing that often displaces poorer residents and results in an influx of more affluent people in a neighbourhood (i.e. gentrification). The focus is typically on safety and security (as testified by indicators such as the safety and security index of Rotterdam) and on efforts to increase the level of (experienced) social cohesion (e.g. through the support for street barbecues). Not only the constant questioning of the effectiveness of such policy- and market-based interventions, also the economic crisis and the accompanying budget cuts put these kinds of interventions under severe pressure. In addition, the Netherlands is seeing the surge of the "participation society", ideas related to the "Big Society" in the United Kingdom. This discourse is accompanied with a withdrawal of the welfare state and severe budget cuts. In Carnisse, specifically, old welfare policy measures were dismantled and specifically, support of public facilities, such as community centres, were abandoned.

Alongside this context, there is an increasing trend of self-organised and bottom-up initiatives by local communities and social entrepreneurs in the Netherlands (Hajer, 2011). This trend is often incorporated in the dominant framing of the "participation society", a withdrawal of the state and dismantling of welfare structures (see introduction). This combination of discourses leads to the following challenge: in areas where the level of self-organisation among local communities is perceived as the lowest, the demand for self-organisation in order to tackle multiple systemic problems becomes the highest (i.e. the more extensive and complex the challenges, the higher the demanded for self-organisation and self-resolving capacities). In addressing this challenge, the public sector is increasingly looking for innovative modes of governance.

Veerkracht Carnisse is rooted in both traditions as these current debates explicated above: all of the four partners in the consortium are relatively young organisations (the oldest was established in 2002) and have been brought up in them. Having been active in different neighborhoods and districts in Rotterdam, they all identified and encountered systemic flaws that led to occasional contact and sporadic short-term collaborations. From this, the idea of a more extensive and longer-term collaboration grew, since the organisations recognised themselves in each other's societal critique, vision and practices.

4.2. Framing its distinctiveness

It was against this backdrop that practitioners were asked to experiment with social innovation in Carnisse within a "living lab" named Veerkracht Lab. In 2009, the concept development and lobby with the municipality of Rotterdam began, which proved to be an enduring process due to a lot of contestation from different municipal department and policy makers. In 2011, the program was finally approved and it eventually started in September 2011, with a running time of four years. The approval for the Lab took two years due to the self-framing of the partners within Veerkracht. The Veerkracht Lab is a vivid example of the contradictions and the fluidity of urban transitions. The clash of paradigms and principles between the different partners involved in the Lab is apparent: inclusiveness, cooperation, network-structure, community-based and organic development contrasts with controlling, accountability, hierarchy and repression.

The Veerkracht Lab aimed to identify new ways of neighborhood development by questioning local democracy, governance and power relations; by the establishment of new networks and by performing self-proclaimed innovating practices.

The underlying question of the program was: what is the transformative capacity of social innovation on the scale of an urban (deprived) neighborhood? In answering this question Veerkracht Lab tries to explore pertinent societal questions when it comes to the more social aspect of sustainability and connections with broader debates around social cohesion, inclusiveness, empowerment and civic rights. What ties the four different partners is the need for a "regime-shift" or fundamental change(s) in not only their own field of interests but in society as a whole.

Confronted with accumulating social-economic problems, Carnisse has been the target of numerous programs by national and local governments for improving housing, security, schooling and working (City of Rotterdam, 2009, 2011; Programmabureau NPRZ, 2014). These efforts are interwoven with broader developments, such as the economic crisis, budget cuts, a reforming welfare state and calls for a "participation society" (comparable to the UK discourse on "Big Society", Bailey and Pill, 2011; Jordan, 2011). Rather than isolated issues, the problems in Carnisse are interlinked and could be argued to be of a persistent nature. The Veerkracht consortium is critical on the current policy that tries to address these problems. They claim that dominant actors nervously cling on to the old way of thinking and developing: they combine top-down spatial-economic revitalisation strategies with a call for "active citizenship" as a response to severe budget cuts. They were weary of the number of short-term and fragmented interventions that seem to lack a more long-lasting effect and sustainable character. As such, these projects only offer a short-term empowerment to the neighborhood, do not build up on precursor programs nor projects and in this way do not ensure the continuity of the learning processes set in place in the area.

Participants in the Veerkracht Lab state that their aim conflicts with this, since the focus is much more on integral collaboration (between the different fields) and social innovation and the lab had to run for at least four years (which was one of the conditions). They claim that the Lab tries to connect to an alternative discourse within urban planning in the Netherlands. This discourse focuses on a proclaimed growing social movement of inhabitants and social entrepreneurs that tries to reclaim public spaces and engages in innovative practices in the urban public sphere (for instance urban gardening, community bonds, local currencies, co-creation of public squares and the self-maintenance of community buildings). These niche trends are based on alternative paradigms and other principles and – according to the Veerkracht-partners – appear to be better suited to current civic demands and socio-economic and ecological circumstances.

In trying to distinguish the Lab from the status quo in neighborhood development, Veerkracht Lab drew up five guiding operating principles of the Veerkracht Lab during several monitoring sessions in 2012–2013 (monitoringsrapport 2011–2013):

- Strengthening and utilising the self-organising capacity of its people foster the resilience of Carnisse. It is about talking with people instead of talking about people.

- People and places are the starting point. From here "learning infrastructures" are built, guided by the daily routines and lifelines of individuals and their networks.
- Methods and activities are developed in an organic manner in the Lab to fit the needs of the area, that is from a practical and operational logic.
- Connections and collaborations (on several levels) are seeked based on innovation and reciprocity. This implies less pressure of bureaucratic control, rules and procedures.
- Look for a balance between top-down intervention and bottom-up self-organisation.

The different partners carry out these principles, although every partner has its own expertise, assumptions and methods. The common denominator is "the emancipatory nature" in regard to the target groups, whether this is a parent, a child, an engaged resident or a volunteer.

4.3. Setting the scene

The choice to land in Carnisse was rather arbitrary. Because Carnisse has a relatively large private housing sector in relation to other comparable neighbourhoods (about 85 per cent), it is atypical in the sense that one of the dominant strategies (demolishing old houses for new houses) and investors (housing associations) are not applicable to Carnisse. It was therefore decided within the municipality that Carnisse would be a fitting place, because "there was not really something going on". None of the partners had been active in Carnisse before 2011, although they were active in the larger district of Charlois. Because of this lack of local knowledge the Veerkracht participants tried to infiltrate the local networks by doing interviews (by engaged researchers and television makers) and exploring collaborations with volunteer organisations, local shops, welfare organisations, etc.

The backdrop of such an institutional setting, a long history of policy efforts and participatory processes left inhabitants weary of "outside" involvement. Such distrust is also linked to these projects' tendency to portray the neighbourhood as disadvantaged, an image with which frustrated many locals and in which they do not recognise themselves. Along with the budget cuts, changing responsibilities of government and citizens as well as the erosion of old welfare structures outlined above, this proofed a challenging context to frame and start a living lab. Another challenging feature that participants raised was the relative openness of both the process and outcome, what led to a certain amount of scepticism by local policy makers and inhabitants on the outcomes of the Lab process.

4.4. Contextualisation of the lab

Fast forwarding to 2015, it becomes apparent that the Veerkracht Lab is contextualised to the dynamics and characteristics of Carnisse: it is active on several primary schools, community gardens, neighbourhood centre's and families that are routed in the social and institutional fabric of the neighbourhood and it's local communities. The primary focus has always been the official administrative borders of Carnisse and this was advocated and commanded by the local commissioner, the sub-municipality of Charlois. In practice, these borders proved rather fluid: for example children at primary schools in Carnisse lived in other neighborhoods and districts, and volunteers at the garden sometimes lived in other cities or villages nearby Rotterdam.

The contextualisation of the process was a notable feature, since it was also one of the five key principles of the lab ("develop and contextualize methods and activities in an organic manner, i.e. from a practical and operational logic"). As stated before, the partners and participants involved in the Veerkracht Lab do not explicitly call themselves an "urban living laboratory" or a "lab". Yet, they do talk about experimenting (an experimental program) and testing new methods and practices. The main element of experimentation is the application of the different methodologies in other contexts, namely:

- Home: Parental coaching and children's coaching.
- School: Physical integrity.
- Outdoors: Neighbourhood gardener.
- Neighbourhood: Transition management and community arena.

The Lab was not only about contextualising existing methodologies; it was rather about further developing existing and new methodologies. All of the methodologies mentioned above were rather abstract visions, instead of well founded in practice. Indeed, some were actually developed and evolved over the course of the Veerkracht-program, such as the Community Arena where a vision of Carnisse of 2030 was formulated, the neighbourhood gardener and the coaching programs. In 2011, some parts if the program were nonexistent or still very fragmented, like the coaching programs at home were fragmented over five different activities (Learning Together, Amigo's, Wegwijs in Nederland, Healthy Lifestyles, etc.). This again confirms the notion above that innovation in Veerkracht was more a discursive act in 2011 and progressively shifted to be a practice-based innovation. In sum, the program gave the partners an opportunity to test their methodologies for urban regeneration and further develop them.

Next to these methodologies Veerkracht Lab experimented with the management and maintenance of public facilities. An example of this is the local community centre, which was shut down in 2011 due to the bankruptcy of the local welfare organisation and budget cuts from the municipality. Veerkracht assisted the local residents and volunteers in organising a process to reclaim the community centre and run it in a self-sufficient manner. Another example is the Carnissetuin, which was a professionally led educational garden that was shut down in 2012 due to budget cuts from the local municipality. Veerkracht assisted the local residents and volunteers in transforming this closed off garden into a community urban garden. Reclaiming these public facilities and central meeting places in Carnisse was a technique to create symbols for resistance and transition.

Next to these experiments aimed at transforming the public places and facilities, new activities were added to the portfolio of the Veerkracht Lab. Examples are the "Moeders van Rotterdam" that focused on assisting and supporting single-parent mothers with new-borns and urban agriculture that focused on growing crops and herbs with volunteers and school kids on a relatively large scale. But next to the development of new activities and practices, others were cancelled, for example the "Taalmap" (teaching the Dutch-language) and the "Gezonde leefstijl" (healthy lifestyles) both aimed at families in Carnisse. Another dimension that was tested, were conceptual ideas, like the Community Bond, a funding scheme for "the local commons" in order to try to embed financial flows more within the neighbourhood and local communities, instead of these financial flows running through them without any adherence or durable effect.

However, playing into the local dynamics and intricacies proved to be challenging due to the specific problems of the neighbourhood such as the erosion of institutional and social networks and a developing atmosphere of mistrust and competition which were a result of cut backs in public and private funding and a history of policy interventions from outside-actors. While some of the partners stated in retrospect that they forced their "contextualised" views and methods on certain stakeholders in the neighborhood, others took a more cautious approach, due to the weariness with outside projects from residents and professionals.

4.5. Participation and collaboration

The competitive sphere led to a rather fragmented approach in the beginning of the program. Every partner went their own way and focused on setting up and contextualising their key activities. Because certain figures in the consortium (at the strategic level) often expressed this competition in a discursively violent way, this complicated the internal collaboration in the beginning of the program. However, by each year there had been a significant increase in the synergies within the consortium and the activities. But in dismissing "the other" and telling other organisations in Carnisse how to do their work, other actors in Carnisse were easily suspicious in collaborating with the different partners or were even offended by them. In some cases this lack of trust and/or aversion could be over won by a respectful tone of voice by operational members of the Veerkracht program, even though in some cases the aversion remained the same.

The arrival of Veerkracht in 2011 came at a salient moment in Carnisse. Against the background of budget cuts and failures of previous participatory processes in Carnisse, the framing of Veerkracht proved difficult. Why did this consortium receive money while the inhabitants were confronted with budget cuts everywhere? They had just seen two of their community centres being closed down and several of their welfare and youth workers laid-off. Expectations with regard to the work of the Veerkracht consortium were critically scrutinised.

As stated before ("setting the scene"), the open nature of program (in terms of outcome) led to a certain amount of scepticism by local policy makers. Other barriers in the process of getting the Veerkracht program approved and started were the bureaucratic accountability relations (e.g. conflicting interests) and the high fluctuation among decision makers. Another difficulty participants raised was the engagement of people in marginalised positions and people with diverse cultural backgrounds.

Starting up the core activities helped making Veerkracht more visible in the neighbourhood and gained the trust of the local communities, though the sphere of competition and mistrust was still tangible. Some activities where seen as a sort of add-on to current activities in the neighbourhood, e.g. a participatory process focusing on the future quality of life of Carnisse (the Community Arena) and primary schools activities. Other activities were more welcomed such as an intervention supporting local change-agents to re-open the community centre, or the reopening of the Carnissetuin in a cooperative manner.

Physical and public places proved to be central in promoting collaboration. It was here that collaboration was put into practice. Central places were in the Veerkracht Lab were the two gardens in Carnisse, the three primary schools in Carnisse, the community centre, residential homes of families in the Frontlijn program and residential streets where activities were organised. It was in these places were the different target groups interacted and collaborated in shared activities.

Most of these collaborations were fuelled by a mutual interest in working together as to reach corresponding goals. For an organisation as Creatief Beheer this mutual interest proved to be crucial in order to invest in longer-term collaboration. As respondents state that these collaborations are built on "reciprocity" and "trust". In the four years of Veerkracht there were different types of collaborations, for example: short- and long-term, incidental and structural, one-sided (or "parasitic") and reciprocating. Some collaboration did not last too long, as others were growing more stable and structure was built in. It proved to be a challenge for all of the partners to create structural collaborations with other actors that were active in Carnisse (e.g. welfare work, religious institutions, etc.). In answering the interviews, respondents questioned the durability or sustainability of the collaborations were, for example primary school teachers, volunteers or community workers able to continue the activities on their own and with their own funding (when Veerkracht would leave in September 2015).

5. Conclusion

The Veerkracht Lab shows how far urban experimentation contextualises its design, processes and practices to characteristics of the district in order to be innovative and tap into the ever changing dynamics. Practitioners try to infiltrate the district and its communities by being present in diverse meeting-places (like schools, community centres, and in public spaces) and thus engaging with local residents, civil servants and entrepreneurs. Based on the findings Veerkracht Lab develops tailor-made interventions based on specific district and community dynamics, while also holding on to generalised methods and practices. In doing so, Veerkracht Lab tries to increase the effects of its practices and make these more durable (especially in a neighborhood like Carnisse).

When we focus on the different translation techniques, the lab shows that the local entanglement and framing the distinctiveness are related to broader debates and developments in Rotterdam and the Netherlands (and even abroad). The lab is connected to and framed in relation to diverse neighborhood discourses and waves of neighborhood interventions. However, the other translation techniques, contextualisation, setting the scene and the participation and collaboration are much more locally embedded in the context of Carnisse. Place is thus a layered notion in the experimentation of the lab, since there is a constant zooming in and out occurring in the Labs' activities and semantics. The Lab taps into the local context, but it also feeds itself with the apparent narratives, symbols and networks.

In terms of the narratives of place, the experimentation process in the Veerkracht Lab created a new narrative of change. This was summarised in the form of the operating guiding principles. The focus on the connection of people and places as "the starting point" for "learning infrastructures" in the area reveals the shift from a reductionist understanding of place as infrastructure to its understanding as a socially mediating facility for change and development. In all the operating guiding principles the notion of collaborative governance pertains, showing that for escaping stigmatisation of the place and its people, reciprocity and institutional connections are key. In this way, the Veerkracht Lab, and the narrative of change co-created in it, establishes a connection between the context of urban regeneration processes at the city level and its local social innovation processes in re-establishing links between place and people.

The symbolic meanings of Carnisse were also an aspect the Veerkracht Lab tapped into, mainly by focusing on public facilities that were closed down due to budget cuts. The different places that were re-established with a different objective, establish a symbolic meaning in the area, being symbols of the resilience of its citizens. But they also were symbols of broader societal change (e.g. "the Participation Society"), and of resistance (against budget-cuts, stigmatisation and outside involvement of professionals). Also, most of the activities in the Lab were focused on empowerment and learning and were experienced by participants as such. This also fuelled the narrative of change and resilience.

In terms of the relations and networks, the activities and public facilities were experienced as locus of meeting each other and enhancing social capital. The Hart voor Carnisse and the Carnissetuin re-established places of encounter in the neighborhood and set up bottom-up collective practices of transforming the relations between residents and between their place and their everyday routines. Forging these collaborations in places like Carnisse is a delicate process, due to the sceptical and cynical views on innovation and weariness with outside interventions. So, it is important to consider that impact of ULL cannot be showcased nor ripped on a short term after their establishment, given that social innovation requires entry to networks, understanding and intervening in social relations and dynamics that are time lengthy processes.

Returning to our central question: in what way does place-embeddedness influence the design, practices and processes of ULL as experimentation settings for urban sustainability transitions? While the Veerkracht Lab formulated principles based on its entanglement within local and national discourses and legitimised itself by formulating its distinctiveness against the backdrop of these discourses, the Lab had to redesign itself and adapt its practices and processes to the context of Carnisse and its institutions, networks and symbolic meanings. Our case study shows that a lab can be "dropped" in a certain place (as Veerkracht in Carnisse) but then the actual design, practices and processes are context-depended. The translation techniques proved to be a necessity for being accepted and being able to effectively function in the local context. However, the Veerkracht Lab also created a sense of place that transformed the urban neighborhood via strengthening the ties of its community to the place. Its activities strengthened the topophilia and thus the lab also became an unintended process of place making.

By addressing the notion of place-embeddedness we tried to address different kinds of impacts of urban experimentation. In doing so we also open up an array of further discussions and research questions as well as the necessity for more comparative research on place-based urban experimentation. It is for instance interesting to see whether ULL have different degrees in their level of place-embeddedness and what this means for their impact and functionalities. In how far do these different place-embedded ULL contribute to addressing sustainability transitions in a certain context?

Notes

1   Researchers set up a monitoring framework with a methodology in combination with aims, indicators and guiding questions for the monitoring. Activities in this monitoring process were writing a yearly progress report and evaluating the Veerkracht Lab as a whole. In order to collect data all partners in the consortium had to deliver their results each year in a quantitative way (e.g. how many people participated in which activity). Also, a total of 158 qualitative interviews were held in a period of September 2011– August 2015. In this same period numerous participatory observations and field visits were carried out, as well as monitoring sessions with different involved actors and an extensive desk study (collecting academic and grey literature in order to grasp the discourse of current debates in welfare reform and neighbourhood development). The result is a rich set of data that relate to different perspectives on life in Carnisse, for example from inhabitants, policy makers, funding agencies to actors involved in the living laboratory.

2   www.urbanlivinglabs.net/

3   In the interviews, we focused on several categories of questions: personal questions and field of interest, questions about Carnisse, questions on Veerkracht-practices, questions on collaborations and questions about specific collaborations with the Veerkracht Lab. The monitoring sessions were aimed at identifying lessons learned from the past years. The participatory workshop the process and impact of Veerkracht was discussed with 25 people from all actor categories and revolved around the durability/sustainability of

the interventions and practices from Veerkracht. In the monthly progress meetings the Veerkracht consortium exchanged updates and activities in the field within an informal setting (about five people present per meeting). In the numerous amount of (participatory) observations and numerous field visits the aim was to grasp the different practices within Veerkracht, by attending, for example judo classes at primary schools, cooking clinics on the garden, educational workshops for volunteers, political debates about the community centre and garden, neighbourhood lunches, etc. These visits and observations were complementary to the other research activities in order to get a full scope of the design, practices and processes of the Veerkracht Lab.

4 Carnisse has the single lowest average income to spend per year in Rotterdam (€ 23,300 in 2014), it has a relatively old and neglected housing stock, there is a lot of mobility and migration streams in the neighborhood (approximately 55 per cent of the people live less than five years in the neighborhood). Carnisse scores low on different municipal indexes (in regards to safety, social cohesion and housing).

References

Binz, C., Truffer, B. and Coenen, L. (2014). 'Why space matters in technological innovation systems – Mapping global knowledge dynamics of membrane bioreactor technology'. *Research Policy, 43*(1), 138–155.

Castán Broto, V. and Bulkeley, H. (2014). 'Maintaining experiments and the material agency of the urban'. In S. Graham and C. McFarlane (Eds.), *Infrastructural lives: Urban infrastructure in context.* London and New York: Earthscan from Routledge.

Coenen, L., Benneworth, P. and Truffer, B. (2012). 'Toward a spatial perspective on sustainability transitions'. *Research Policy, 41*(6), 968–979.

Edwards-Schachter, M.E., Matti, C.E. and Alcántara, E. (2012). 'Fostering quality of life through social innovation: A living lab methodology study case'. *Review of Policy Research, 29*(6), 672–692.

Entrikin, J.N. and Tepple, J.H. (2006). 'Humanism and democratic place making'. In S.Aitkin and G. Valentine (Eds.), *Approaches to human geography* (pp. 30–41). Thousand Oaks, CA: Sage.

Frantzeskaki, N., Loorbach, D. and Meadowcroft, J. (2012). 'Governing transitions to sustainability: Transition management as a governance approach towards pursuing sustainability'. *International Journal of Sustainable Development, 15*(5), 19–36.

Frantzeskaki, N., Wittmayer, J. and Loorbach, D. (2014). 'The role of partnerships in "realizing" urban sustainability in Rotterdam's City Ports Area, The Netherlands'. *Journal of Cleaner Production, 65*, 406–417. Retrieved from http://dx.doi.org/10.1016/j.jclepro.2013.09.023

Franz, Y. (2015). 'Designing social living labs in urban research'. *Info, 17*(4), 53–66.

Gieryn, T. F. (2000). 'A space for place in sociology'. *Annual Review of Sociology, 26,* 463–496.

Gieseking, J. J., Mangold, W., Katz, C., Low, S. and Saegert, S. (2014). *The people, place, and space reader.* London: Routledge.

Hajer, M. (2011). *De energieke samenleving. Op zoek naar een sturingsfilosofie voor een schone economie.* The Hague, The Netherlands: Planbureau voor de Leefomgeving.

Kisby, B. (2010). 'The Big Society: Power to the people?' *The Political Quarterly, 81*(4), 484–491.

Lehmann, V., Frangioni, M. and Dubé, P. (2015). 'Living Lab as knowledge system: An actual approach for managing urban service projects?' *Journal of Knowledge Management, 19*(5), 1087–1107. Retrieved from doi:10.1108/JKM-02-2015-0058.

Liedtke, C., Welfens, M., Rohn, H. and Nordmann, J. (2012). 'LIVING LAB: User-driven innovation for sustainability'. *International Journal of Sustainability in Higher Education*, *13*(2), 106–118.

Ministry of Housing, Spatial Planning and the Environment. (2007). *Actieplan krachtwijken. Van aandachtswijk naar krachtwijk*. The Hague, The Netherlands: Author.

Pereira, L., Karpouzoglou, T., Doshi, S. and Frantzeskaki, N. (2015). 'Organising a safe space for navigating social-ecological transformations to sustainability'. *International Journal of Environmental Research and Public Health*, *12*, 6027–6044. Retrieved from doi:10.3390/ ijerph12060602.

Putters, K. (2014). *Rijk geschakeerd. Op weg naar de participatiesamenleving*. The Hague, The Netherlands: Sociaal en Cultureel Planbureau.

Ransome, P. (2011). ' "The big society" fact or fiction? A sociological critique'. *Sociological Research Online*, *16*(2), 18.

Smith, A. and Raven, R. (2012). 'What is protective space? Reconsidering niches in transitions to sustainability'. *Research Policy*, *41*, 1025–1036.

Stedman, R.C. (2004). 'Is it really just a social construction? The contribution of the physical environment to sense of place'. *Society and Natural Resources*, *16*, 671–685.

Stedman, R.C. and Ingalls, M. (2014). 'Topophilia, biophilia and greening in the red zone'. In K.G. Tidball and M.E. Krasny (Eds.), *Greening in the Red Zone: Disaster, Resilience, and Community Greening*. New York: Springer-Verlag.

Steenbergen, F. and Wittmayer, J. (2012). *Carnisse in transitie?* Erasmus Universiteit Rotterdam, The Netherlands: DRIFT.

Tidball, K. and Stedman, R. (2012). 'Positive dependency and virtuous cycles: From resource dependence to resilience in urban social-ecological systems'. *Ecological Economics*, *86*, 292–299.

Tomaney, J. (2014). Region and place II: Belonging. Progress in Human Geography. Retrieved from http://dx.doi.org/10.1177/0309132514539210

Tonkens, E. (2014). Vijf misvattingen over de participatiesamenleving. Afscheidsrede Universiteit van Amsterdam, The Netherlands.

Visitatiecommissie Wijkenaanpak. (2011). *Toekomst van de wijkenaanpak: doorzetten en loslaten* (Deel 1). The Hague, The Netherlands: Visitatiecommissie Wijkenaanpak.

Voytenko, Y., McCormick, K., Evans, J. and Schliwa, G. (2015). 'Urban living labs for sustainability and low carbon cities in Europe: Towards a research agenda'. *Journal of Cleaner Production*, *123*, 45–54. Retrieved from doi:10.1016/j.jclepro.2015.08.053.

Wilbanks, T.J. (2015). 'Putting "place" in a multiscale context: Perspective from the sustainability sciences'. *Environmental Science and Policy*, *53*, 70–79.

Williams, D.R. (2014). 'Making sense of "place": Reflections on pluralism and positionality in place research'. *Landscape and Urban Planning*, *131*, 74–82.

Williams, D.R. and Stewart, S.I. (1998). 'Sense of place: An elusive concept that is finding home in ecosystem management'. *Journal of Forestry*, *96*(5), 18–23.

14

CONCLUSIONS

Simon Marvin, Harriet Bulkeley, Lindsay Mai,
Kes McCormick and Yuliya Voytenko Palgan

1. Introduction

This book has sought to advance the theorisation of the interface between experimentation, socio-technical transitions and the city, through an internationally comparative empirical analysis of ULL. This has been structured through three core elements. First, through a set of conceptually informed empirical case studies from Africa, Asia, Europe and North America, based on original empirical work through case studies and interviews with municipal staff, private sector representatives, activists and civic communities, technologists, developers and others involved in the making of ULL. Second, by undertaking a critical examination of the discourses and practices associated with ULL, including an evaluation of what new capabilities are being created by whom and with what exclusions; how these are being developed and contested; where this is happening both within and between cities and, with what sorts of socio-economic, political, environmental and material consequences. Third, by developing an original framework informed by the body of work on transitions theory and work on urban governance and politics. The risk of such an approach is theoretical eclecticism. To counter this risk, the framework focused on a set of core concerns across different approaches and sought to identify the common ground they share: the design, practices and processes of ULL. In developing this concluding chapter, we follow the structure of the book and consider how our comparative, critical analysis of ULL provides new insights into their design and practice, and the processes through which they come to be embedded in sustainability transitions. In each section, we summarise the main findings, examining how the chapters address the core questions that underpin each part. In our final section, we review the wider policy and research implications of the book setting out key future priorities.

2. The design of ULL

The first part of the book focused on understanding the ways in which ULL are being designed and how they vary between urban contexts. ULL are examined in the following countries: the United Kingdom, Sweden, Germany, The Netherlands, Austria, Finland and the United States. The chapters discuss a set of ULL design components including: goals and visions; actor constellations and agency; actor engagement in ULL set up; implementation and evaluation; the themes and topics that ULL engage with; funding sources and the placement and geographical embeddedness of ULL. The sheer variety of elements here suggests that designing ULL is not a straightforward matter of developing a formulaic plan or blueprint, but is rather an on-going project.

Focusing on the work involved in the design of ULL reveals the importance of forming coalitions and intermediary institutions, establishing shared understanding of challenges of sustainability transitions in relation to particular contexts for intervention, identifying the technological interventions to be trialled and agreeing on the governance principles to be followed. Each set of decisions is open to contestation, and the chapters reveal that the on-going work required in order to embed and establish an ULL is considerable. The critical questions asked in the first part of the book focused on three questions: first, how the visions, institutions, intermediaries and technologies required to deliver ULL are assembled and embedded; second, what principles and modes of governance are being developed through ULL and third, why and how the design of ULL affects the practices they deploy, their effectiveness and the implications for their ability to foster sustainability transitions. An important contribution of the chapters in this first part of the book has been also to examine how and why all of the above ULL design attributes vary across urban contexts. Taking each of these questions in turn, we reflect on the key findings by looking across the chapters in this part.

First, the issue of assembling and embedding the capacity to organise a ULL. Voytenko Palgan and colleagues provide an informative analysis of the landscape of ULL and how they are being designed as a form of collective governance and experimentation across different urban contexts. Their two central findings are as follows: first, the ways in which ULL is a mode of experimentation that seeks to work across different urban imaginaries of the smart, low carbon and resilient city, and second, the way in which the technique has resonances and dissonances with established modes of neo-liberal urban governance primarily focused on competitiveness and economic growth. Consequently, they point to the consistency of ULL with existing economic partnerships and traditional urban priorities that are brought together in what are claimed to be new ways – more open, collaborative and experimental ways of "doing" urban development. This analysis provides us with a much more modest and critical framework for thinking through the distinctiveness and potential of ULL.

Second, the issue of how the mode of governance shapes the design and priorities of the ULL. The next two chapters by Hodson and colleagues and

Levenda provide conceptually informed and empirically rich examples of the interface between existing social context modes of organisation and urban priorities and the claims of the distinctiveness of ULL. Hodson et al. focus on cycling experimentation in greater Manchester, which provides a fascinating case of "constrained experimentation". Two decades of using the city region as a test bed for national priorities and initiatives meant that what was regarded as an experimental process was highly constrained by prior and pre-existing assumptions and commitments at both the national and city-regional levels. The process of experimentation had weak coordinated capacity and although many of the city's regional participants learnt about the limits and potentials of cycling these were learnt through systemic process and did not usually inform decision making. There is clearly a need to develop processes and politics that create space for users and more effective learning that can work beyond the limits of existing state and city-regional constraints. Similarly, Levenda in the chapter on smart gird experimentation in the United States illustrates how an experimental process was framed as a form of technological testing and an economic opportunity for new smart grid products and services. Within this framework, the role of users has been highly prescribed focused on consumption habits and technology adoption – in which users were seen as barriers to smart grid implementation. Both these very different contexts and infrastructures of experimentation graphically show how existing context effectively constrains the boundaries of experimentation. Each has a limited concept of how users can interact and participate in a sustainability transition and point to the need to reinvigorated experimentation with a wider range of options and voices.

Finally, all the chapters in Part I raise difficult questions for wider debates and analysis of the role of ULL in promoting or stimulating sustainability transitions in the socio-technical organisation of urban infrastructure. They show the primary focus of techno-economic innovation as urban priority and appear to be less transformative than the claims made. Existing priorities and mode of organisation are more important than transformation. Sengers et al. critically examine these issues in more detail and ask how we can conceive of the relationship between programmes of experimentation in ULL and sustainability transitions. They make two key points to understand this relationship. First, that we need to be careful to assume that claims of transformation around ULL may not actually result in visions that are guided by a future of a very different society or is necessarily inclusive of actors from outside established policy processes. They argue that it is important to critically examine ULL to assess the extent to which these are practices designed to maintain business as usual or whether they imply some profound transformation. Second, they point to the difficulty of assessing whether a transition has taken place because of the multiple and interacting relations between different socio-technical systems, for example the interconnection between smart and energy implies new forms of hybridity and entanglement. This makes it much more difficult to assess whether a transition has occurred.

3. The practice of ULL

In the second part of this book, contributing chapters illustrate how the practice of ULL may involve different forms and techniques for learning, shielding, nurturing, empowering and participating within ULL. The chapters develop an analysis of the everyday work of ULL and their impacts in particular urban contexts focusing on the three sets of questions: first, niche practices of learning, shielding, nurturing and empowering within ULL, who is undertaking these practices; second, which practices appear to be more or less successful, and why and, third, how more effective practices can be designed to develop ULL in different urban contexts. Together they suggest that what it is that ULL can accomplish is as much derived from the on-going and often mundane practices involved in their realisation as it is "written into" their initial design. They point to the ways in which ULL may evolve over time, as forms of social and material resistance are encountered, negotiated and new practices initiated. We examine below how each chapter deals with these questions.

Davies and Swilling present the foundations of collaborative governance in Stellenbosch, South Africa, to highlight the specific contextual conditions that allowed for the emergence and evolution of a dynamic ULL. Critical to this was a collective commitment by the academy and partners to practicing transdisciplinary research and pioneering a unique approach to research partnerships that support the conduct of scientific research with rather than for society opening up novel opportunities for innovation and experimentation. This required a process of breaking away from traditional "extractive" research modes that define researchers as the generators of solutions and societal actors as the consumers of these solutions. As argued in this chapter, the effectiveness of the response requires researchers to develop new skills as process facilitators and to appreciate abductive modes of inquiry that allow problem statements to be jointly formulated by researchers and societal actors. This chapter demonstrates that this shift is possible if emergent transdisciplinary design processes are appropriately facilitated to move away from systems knowledge about "what exists" and target knowledge about "what should exist", to transformation knowledge about "how things can change".

Astbury and Bulkeley focus on Manor House PACT – a project that sought to develop community-based and community-led responses to urban sustainability challenges. This chapter introduces the work of PACT and reflects on the ways in which it served to constitute a form of ULL in which a place-based community became the arena within which multiple approaches to sustainability designed to engage excluded communities were developed and tested. Seen from this perspective, they show how PACT was a distinctive form of ULL that served both as an enclave of sustainability and as a test-bed for new solutions. But rather than operating from the "top down", the capacity of PACT to govern is conditioned through an array of techniques that work iteratively, adjusting to experiences and adapting the forms of intervention working with different constituents in real-time. The chapter focused on the processes through which PACT has sought to extend its influence beyond its boundaries, particularly focusing on the ways in which the constitution of PACT as part of a wider programme of community initiatives has enabled it to circulate, reaching into arenas, and how its approach to learning has been central in fostering its transformative potential. In this way, Manor House PACT has functioned explicitly as a laboratory for processes of translating, learning, scaling and empowering.

Davies interrogates the experience of co-designing and conducting a suite of HomeLabs experiments in Dublin, Ireland seeking to provoke transitions to more sustainable household consumption. Developed with a commitment to co-design principles, the HomeLabs embody a research-led, experimental living laboratory approach. Adopting a social practice orientation and building on a participatory backcasting and transition framework approach, the HomeLabs focused on combining, aligning and testing promising tools, rules, skills and understandings for more sustainable and low carbon consumption. Focusing specifically on the experiences and outcomes of the washing HomeLabs for the participating households and the wider governing environment, this chapter reflects on the opportunities and challenges of such techniques for domestic sustainability transitions under conditions of austerity. As such, it highlights the importance of context when initiating discussions around comparative living laboratory endeavours and brings to the fore empirical evidence of experimentation for sustainable transitions during a period of dynamic regime change rather than stability. In the absence of direct and explicit strategic niche management for sustainability transitions from the national government, experimentation through HomeLabs fits the model of a less directed process with diverse actors leading actions to stimulate new spaces of innovation for the testing and evaluation of alternative practices.

Finally, Castan Broto investigates the extent to which smart city visions are fit to deliver both environmental sustainability and social justice in contemporary cities. This is approached through a retrofitting perspective, which engages with the potential to improve service delivery in existing cities leaving aside "fantasies" about the possibility of creating new smart cities from scratch. This illustrates how urban redevelopment projects are never developed on a blank slate as future urbanisation depends on the transformation of existing land use relations. While there is a promise embedded in the idea of experimenting with smart cities the chapter illustrates that there are clear challenges for attaining cities that are socially and environmentally just. Thinking about the creation of operational and citizen-oriented smart cities will depend on the extent to which smart city visions, and the means deployed to deliver them, respond to the needs of cities and their material constraints. The analysis follows three case studies in Hong Kong, Bangalore and Maputo and shows, first, the heterogeneity of perspectives that emerge to characterise smart cities, and, second, their realisation as an on-going project of urban development which is never really accomplished.

4. The processes of ULL

Understanding the processes through which ULL create an impact beyond their immediate domain is critical in terms of assessing their potential as a means of governing transitions for sustainability. There are three key questions: first, how, why and by whom ULL innovations are translated into urban socio-technical systems and regimes; second, which mechanisms are most effective in supporting the translation and up scaling of ULL into urban governance domains and third, how the design, implementation and practice of ULL can be improved to support the impact and effectiveness of ULL. Together, the chapters in this part of the book, and the contributions across this volume, show that ULL are differently disposed to the possibility of achieving change – while some are concerned with transformation of particular places or urban conditions, others seek to create the capacity to engender change beyond their immediate boundaries. We examine below how each chapter addresses these questions.

Trencher and colleagues build on our understanding of potential long-term societal impacts of ULL, and their capacity to trigger broader societal change. The 2,000-Watt Society ULL started with the nationally developed scientific vision and was then scaled down to Basel as a pilot site for applying the vision and testing the concept. Localised projects focusing on building energy efficiency, large-scale urban development and mobility were launched to implement the vision. These projects drew in industrial partners through seed-funding schemes into the university-led ULL engaging mainly with local government. This mechanism was effective in translating the experimental initiative into the broader urban governance context by forming a physical channel to mainstream and raise the visibility of the scientific vision in the public and political domains. Essentially, the design and incorporation of the scientific vision over the long-term created the capacity to bind stakeholders from different sectors of interests into one collective pathway to change that through the regional implementation plan were put into legislation. Yet, the wider impacts of lifestyles on sustainability were disregarded in the original design, which prompted discussion over values for "sufficiency" instead of merely the quantitative amount of energy consumption. Citizens are still not the focus in the anticipated outcomes of the project that stressed technological verification for industries and local government.

In Burch et al. three initiatives were identified for analysis as place-explicit experimental interventions ranging from prototyping to commercialisation and to fostering changes within established businesses. Scalability is the keyword in the Canadian cases focusing on both local and global impacts. Local benefits of these ULL include empowering students in city building, creating green jobs, improving local urban environment and reducing carbon footprints of local businesses. In the meantime, these ULL are also designed with scalable global impacts – delivered through building global identities and responsibilities of students, accelerating commercialisation of innovative products and through developing carbon strategies that are applicable in global supply chains and industries. Although the local municipal government is the core player in initiating, translating and scaling up the initiatives into different urban governance domains, the Canadian cases display complex patterns of multi-scalar and multi-institutional relationships that are considered productive and were intentionally developed by the municipality to enable niche innovations. By opening up access to government assets and issuing progressive policies, the municipal government invited new intermediary actors to catalyse changes and share risks.

Mai examines the characteristics of emerging ULL in China, mapping out the impacts of three different designs and practices of socio-technical innovations. The multiplicities of co-existing experimentation forms of grassroots, civic and strategic natures are starting up dynamic processes of urban transitions with both bottom-up and top-down flows of intentionality, resources, mandates and powers. The collective outcomes of such multiplicities indicate a stage of re-designing existing urbanity rather than a vision of radical transformation. Interventions applied at various scales have resulted in changes of material arrangements, infrastructural settings and management models. Yet, scalability is limited for grassroots initiatives given the geographical constraints of applicability. The intention to translate experimented innovations into other urban contexts and to scale them up to upper levels of governance is discernible in civic and strategic types of interventions but evidence of take-up is limited. Evaluation of innovation outcomes has been instituted in individual initiatives for either funding continuity or for administrative requirements but a learning structure across individual programme evaluation has not been established to build up an understanding collective effectiveness.

Steenbergen and Frantzeskaki employ the concept of "sense of place" to understand ULL embeddedness in a particular socio-spatial context in a relational perspective. Translation techniques for ULL are applied in studying the long-term effectiveness of a community based initiative Veerkracht in a deprived neighbourhood in Rotterdam. New ways of neighbourhood development were devised by questioning local democracy, governance and power relations; by establishing new networks as well as by performing self-proclaimed innovating practices. The transformative capacity of these social innovations in the neighbourhood of Carnisse was explored in connection with debates around social cohesion, inclusiveness, empowerment and civic rights. The stakeholders focus was on fostering regime shift in society as a whole by changing the practices in one neighbourhood to render an alternative approach to the mainstream policy thinking. Self-organising capacity among local people was stressed to build up learning infrastructures with daily routines of individuals and their networks. To enhance collaboration, physical and public places in Carnisse played a central role, in contrast to virtual platforms, demonstrating that collaboration needed to be built on "reciprocity" and "trust".

5. Future societal and research priorities

The issues raised in this book highlight the challenge of determining the societal and research priorities for the future practice and study of urban sustainability transitions. We argue that there are three critical issues to be addressed by both the diverse policy community involved in ULL and shaping their creation as well as the fragmented urban and sustainability research communities interested in understanding their implications and consequences.

5.1. Urban imaginaries: Continued fragmentation or re-invigoration of the sustainable city

ULL clearly bring together, through processes of experimentation, diverse urban imaginaries of the smart city, the low carbon city and the resilient city. The fragmentation of the singular discourse of the sustainable city appears to have been replaced with a more diverse rather than singular, more experimental rather than programmatic and more localist rather than metropolitan wide attempt to develop new urban imaginaries. While the diversity, openness and contingent character of these developments are positive, there are a set of critical questions about what are the wider strategic implications of this fragmented landscape. There are three critical issues to consider. First, we have to consider the extent to which this fragmentation signals a re-intensification of the commitment to a form of ecological modernisation – that the focus should be on responses that are consistent with economic growth. In turn, this may indicate a movement away from any previous commitments to social justice and an enlarged view of urban ecology and environment in earlier concepts of the sustainable city. Second, it is critical to consider what these new attempts to develop a hybrid of smart, low carbon and resilient imaginaries mean for the city. Could such responses be about resolving problems at the lower scale of a building, district or zone or alternatively do they provide a way of thinking about new forms of hybridity at the metropolitan scale – of a means through which the climate smart city can be forged. Third, they raise the question of how we start to explore in a more systemic and programmatic manner the resonances and dissonances between these new experimental logics to better understand what issues become prioritised and what are left or ignored, and with what consequence.

5.2. Organisational modes: Municipal, community and national experiments

There is clearly a diversity of different types of practices in the design and social organisation of ULL. While we have argued, and demonstrated that there are hybrid formations that seek to combine purposes and the social interests involved in ULL there remains a tendency for the separation and demarcation in the design of experiments. Although we acknowledge the potential and diversity of responses, the separation of purposes into market focused commercial driven experiments and the more open and broader objectives of community responses may not actually be helpful in stimulating open and user driven innovation. State sponsored innovation experiments and the propensity of municipal responses to focus on local economic priorities may actually delimit and restrict the boundaries for user led innovation. Paradoxically, the critical importance of state sponsorship through innovation funding and the coalitions between municipal, higher education and corporate actors within ULL may restrict or narrow the roles users can play in the innovation process. Overall there is a tendency for initiatives to undertake forms of experimentation *on* the context and users rather than working co-productively and symmetrically *with* context and users. Consequently, the urban is constituted as a test-bed according to external priorities and processes.

Clearly there is potential for opening the innovation process in the following ways. First, enlarging the concept of innovation to focus less on novel technologies and more on the potential for social and cultural innovations. Such an approach may then create a wider societal context for innovation that might have a techno-logical component. For instance, smart does not have to focus on digital technologies but can also include novel social and cultural practices. Second, to enlarge the notion of innovation itself so it is not solely focused on techno-economic innovations could also include social innovations and innovation within the public and voluntary sectors. There is a tendency to delimit and restrict the concept of innovation and it needs enlarging. Third, rather than focus on short term applications, we need a longer term view of the innovation process. We should not expect the rapid stabilisation and short-term marketisation of these products and services. Earlier rounds of experimentation in the sanitary city and industrial city demonstrate the long periods of contestation, experimentation and testing before infrastructural innovations become socially and institutionally embedded and stabilised. There is a tendency to emphasise the need for speed, rapid innovations rather than longer term processes of incremental innovation and testing – which can take place outside the market context.

5.3. Societal implications: Learning, reflexion and configurational urban transitions

Finally, while the claim is that ULL are about processes of internal learning about the potential for and limits to sustainability transitions the cases indicate that there is significant variability in the capacity of ULL to undertake wider learning about the societal implications of their experiments. Most processes of learning within ULL tend to focus on outcomes and more quantitative indicators. Yet, the processes of second order learning are potentially much more significant for developing active and configurational transitions. This would place a premium on those forms of learning that focus on the wider lessons learnt from the process of experimentation itself rather than a narrower focus on impacts. For instance, what do sets of experiments in ULL tell us about the potential for change in an infrastructural regime – and in particular what are the types of wider changes in the regulation and social organisation of an infrastructure that can create a more amenable context for the successful implementation and potential acceleration of an experimental initiative.

There is a need for a more systemic and programmatic attempt to develop forms of social learning and reflexion across the landscape of ULL that brings these questions into sharper focus and provides ways of linking experience and knowledge across different urban contexts and sectors. There are three implications for future ULL. First, to develop forms of learning embedded in the ULL that do not solely focus on quantitative impacts but can undertake qualitative learning over the life of the ULL. Second, to do so in a more comparative framework across different urban contexts so learning can be co-produced and shared across different ULL in ways that might help build relevant and valid lessons across an urban system. Third, to find ways of interconnecting national innovation systems with the focus on corporate and techno-economic innovation with the messiness and diversity of urban contexts in which innovations are trialled and developed in order to try to shift urban contexts from "test beds" for national experiments to partners in the innovation process.

INDEX

The acronym ULL is used throughout for Urban Living Labs.

Locators in **bold** are used to indicate tables and those in *italics* to indicate figures, though when integrated with continuous discussion in the text these are included in regular page spans.

2,000-Watt Society: analytical framework 169–170; design of ULL 168–169, 183–185; initial implementation 173–178; long-term impacts 15, 168, 172–173, 183, 253; redesign and reconfiguration 178–183; set-up phase 171–173

academia *see* research context; universities
activism in China 211–215
agency, climate change 191–193, 204
Amsterdam as smart city 79, 81, 82
APRILab **23**, 28, 29–32, *30*
austerity conditions 127–128, 133
Austin, Texas, energy transitions 54–58, 59–67
automobile technology 175–176; *see also* transportation

Bangalore, energy landscapes 154–160
Basel Pilot Region laboratory *see* 2,000-Watt Society
Big Lottery 108
buildings, 2,000-Watt Society 174–175, **180**
business *see* commercial context

Canada *see* Vancouver climate change case study
carbon emissions *see* climate change; low carbon cities
Carbon GSP, China 215–220
CASUAL **23**, 29, **30–31**

Celtic Tiger economy 127
characteristics of ULL: in China 225–227, **226**; distinctiveness 2, 7–10, **25**, 26, 238–240; sustainable cities 26–27
charcoal as fuel 156–157
China, emerging ULL in 210–220, 227–228, 254
circular cities 79, 81–82
cities: climate change action 189–190, 192–193, 204–206; design of ULL 12–13; emergence of ULL 6–7; mechanisms for managing 222–227; Rotterdam, the Netherlands, embeddedness case study 235–243; strategic controls 220–222; urban imaginaries 255; *see also* geographical context; low carbon cities; smart cities; sustainable cities
citizen science 151
city of bits 79; *see also* smart cities
CityStudio Vancouver **195–196**, 196–198, 202–204
civic ULL **8**, 9, 215–220
Clean Urban Transport Europe Programme 25
climate change: comparative analysis 201–206; energy transitions 52–53; experimentation in urban spaces 191–194; governance 24, 189–191; socio-technical climate innovators 193, 194–201; *see also* Vancouver climate change case study
Climate Smart **195–196**, 200–201, 202–204

co-design: 2,000-Watt Society 172–173; characteristics of ULL 26–27; domestic practices in Ireland 127–128

collaborative governance, Stellenbosch 91–100

commercial context: climate change action 193, 200–201; European cities 24; smart cities 80–81

Communities Living Sustainably (CLS) programme 108, 110, 119

community context: China's ULL 210–220, 227–228; enabling transformation 106–107, 118–123; governance mechanisms 222–227; Manor House resilience case study 107–118, 251–252; Rotterdam, the Netherlands, embeddedness case study 237–238; strategic controls 220–222

comparative analysis: China's ULL 211, 227–228; climate change 201–206; critical gaps 10; international context 2–3; smart cities 154–160, **160**

CONSENSUS project 128–130, *129*, 141

consequences of ULL 2, 11

conservation in China 212–215

construction: energy efficient buildings 174–175, **180**; literal and figurative 75

consumers, smart grids 65–67

critical gaps in ULL 10–11

cycling in Greater Manchester case study 37–38, 39–49, *44*

design of ULL 11–13, 249–250; 2,000-Watt Society 168–169, 178–185; co-design 26–27, 127–128, 172–173; cycling in Greater Manchester case study 37–38, 39–49, *44*; energy transition in Austin, Texas 54–58, 59–67; smart city experimentation 76–83; sustainable cities 22–26, 27–33

digital cities 79; *see also* smart cities

distinctiveness of ULL 2, 7–10, **25**, 26, 238–240

domestic practices: 2,000-Watt Society 181–183, 185; assembling HomeLabs 128–132; as energy consumers 58, 64–67; in Ireland 127–128; laboratories 126–127; smart cities 151–153; washing practices 130–143; water governance 132–143

eco cities 81, 148–149; *see also* smart cities

economic context: austerity 127–128, 133;

energy transitions 58; *see also* commercial context

embeddedness 231–232, 243–245; longitudinal place-based research 235, 254; ULL as transition sites 232–235; Veerkracht Carnisse, Rotterdam, the Netherlands 235–243; *see also* geographical context

empowerment: CityStudio Vancouver 196–197; enabling transformation 120–121

energy landscapes: Austin, Texas 54–58, 59–67; comparative analysis 154–160; governing through experimentation 54, 58–64; renewable energy 178, **180**; smart cities 29, 153–154; smart consumers 58, 64–67; smart grids 53, 54–58, 60–61; transition 52–54, 67–68; *see also* 2,000-Watt Society

entrapment 4

environmental context: China's ULL 212–215; European cities 29–33; *see also* climate change; energy landscapes

European cities 21–22, 34–35; characteristics of ULL 26–27; comparison to China 210; design of ULL 22–26, 27–33; governance 22–26; knowledge sharing networks 33–34; project examples **23**

European Commission 23–24

evaluation of ULL 8, 27

experimentation: characteristics of ULL 26, 168; climate change action 191–194; cycling in Greater Manchester 46–48; as discursive arena 77–78, 80–82; distinctiveness of ULL 8; embeddedness 231–232; energy transitions 54, 58–64; as institutional reconfiguration 77–78, 82–83; Manor House resilience case study 116–117; as material site 77–80; rationalities of 65–67; smart cities 76–83; Stellenbosch 100–103; sustainability transitions 76–78; sustainable cities 32–34

Federal Institutes of Technology (ETH) 169–173, 178–179

Finland, design of ULL 27

Foucault, Michael 65

funding: China's ULL 225; drivers of ULL 24; Manor House resilience case study 111–112; private sector 28–29

future research 254–257

gardening, urban areas 112–113
Geels, Frank 77–78
geographical context 37–39, 49:
    characteristics of ULL 26; China's ULL
    217–220; climate change action 202;
    cycling in Greater Manchester 37–38,
    39–49; distinctiveness of ULL 8;
    embeddedness 231–232, 243–245,
    253–254; longitudinal place-based
    research 235; smart grid in Austin,
    Texas 57–58; ULL as transition sites
    232–235; Veerkracht Carnisse,
    Rotterdam, the Netherlands 235–243
global context of ULL 2–3
goals: 2,000-Watt Society 176–178;
    climate change 189, 191
governance: city context 38–39; climate
    change 189–191, 192, 194–196; cycling
    in Greater Manchester 48–49; energy
    transitions 54, 58–67; mechanisms of
    managing cities 222–227; smart cities
    152–153, 159–160; Stellenbosch
    91–100; sustainable cities 22–26; urban
    living labs 1–5
Governance of Urban Sustainability
    Transitions (GUST) project 107
grassroots ULL **8**, 9–10; China 211–215,
    222–223; enabling transformation
    106–107, 111, 118, 121–123; smart
    cities 151–152, 159–160; smart grid in
    Austin, Texas 63–64
Green/Blue cities **23**, 29, **31**
Green and Digital Demonstration
    Program (GDDP) **195–196**, 198–200,
    202–204
greenhouse gases see low carbon cities
Green Source, China *212*, 212–215,
    225
Guangdong Province, China 211–220
Guangzhou, China 211–220

Hamburg as smart city 79, 80–81, 82–83
home environment see domestic practices
HomeLabs: assembling 128–132; learning
    from 143–145, 252; water governance
    132–143
homo-economicus 58
Hong Kong, energy landscapes 154–160
households see domestic practices;
    HomeLabs

information cities 79; *see also* smart cities
information and communication
    technology (ICT): China's ULL 215,

219, 227–228; energy landscapes 53, 61;
    European cities 23–24; smart cities
    74–75, 78–84, 147–153, 158
innovation: China's ULL 211–220;
    embeddedness 231–232; emergence of
    ULL 6–7; European cities 23–25;
    governance 3–4; multiples modes of
    ULL 255–256; Stellenbosch 100–103
institutional reconfiguration 77–78,
    82–83
intelligent cities 79; *see also* smart cities
interaction see co-design; participation
international context of ULL 2–3
International Sustainable Campus Network
    (ISCN) 33
Ireland: assembling HomeLabs 128–132;
    domestic conditions in 127–128; water
    governance 132–143
Irish Water 137–139, 143

JPI Urban Europe 22, 28, 33, 107

Kern, Florian 77
knowledge see learning
knowledge partners 28
knowledge sharing networks 33–34

laboratories: community context 106–107,
    122; domestic conditions 126–127;
    see also HomeLabs
leadership: characteristics of ULL 27;
    climate change action 197–198,
    204–205
learning: cycling in Greater Manchester
    48–49; experimentation 26, 33, 34;
    future research 256–257; HomeLabs in
    Ireland 143–145, 252; Manor House
    resilience case study 116–117, 119;
    purpose of ULL 3, 8; smart grid in
    Austin, Texas 60–61; Stellenbosch 93,
    96–101, 103; *see also* participation
living conditions see domestic practices
local governments: 2,000-Watt Society
    171–173, *172*, 183–184; China's
    emerging ULL 214–215, 223, 228;
    climate change 192–193, 201–205;
    design of ULL 28, 29; urban
    governance 38, 59, 253
longitudinal place-based research 235–243
low carbon cities: 2,000-Watt Society 169,
    174–178, 184; China's ULL 215–220;
    European cities 21–22, 29–33; Hong
    Kong 155; *see also* climate change;
    energy landscapes; sustainable cities

Manchester, cycling case study 37–38,
39–49, *44*
Manor House Development Trust
(MHDT) 109–110, 119–120
Maputo, energy landscapes 154–160
mobility *see* transportation
multiple modes of ULL 5–10, 255–256

Nansha, China 211, 220–222, *221*
narratives of place 233; *see also*
embeddedness; geographical context
neighbourhoods *see* community context;
geographical context
The Netherlands: design of ULL 27;
embeddedness case study 235–243
niches: climate change action 204–205;
Manor House resilience case study
111; smart grid in Austin, Texas 63;
strategic niche management 3–4,
144–145, 252

outdoor space, urban areas 112–116, 121
Oxford Road corridor, Greater
Manchester 37–38, 42–49, *44*

PACT (Prepare, Adapt, Connect, Thrive)
107–118, 119–123, 251–252
Paris Agreement 189
participation: 2,000-Watt Society 171–173,
181–183, 185; characteristics of ULL
26–27; citizen science 151;
distinctiveness of ULL 8; domestic
conditions in Ireland 128–132; Manor
House resilience case study 121; the
Netherlands, embeddedness case study
238–240, 242–243; smart grid in Austin,
Texas 63–64; *see also* domestic practices;
grassroots ULL
path dependency 4, 189–191, 194
place *see* embeddedness; geographical
context
placelessness, smart grid in Austin, Texas
57–58
planning, 2,000-Watt Society 174–175,
180
practice-oriented participatory (POP)
activities 127–128
practices of ULL 13–14, 251; collaborative
governance in Stellenbosch 91–103;
comparative energy landscapes 154–160,
160; domestic consumption 127–132,
135–143; Manor House resilience case
study 107–118; smart cities 74–76,
83–85, 148–153, 252

private sector 1, 9; 2,000-Watt Society
174; energy landscapes 62; European
cities 28–29, 33; smart cities 158;
Stellenbosch's governance 91
processes of ULL 14–15, 252–254; 2,000-
Watt Society 168–170, 173–178,
184–185; China's ULL 210–211, 213,
222–228; embeddedness in Rotterdam,
the Netherlands 235–243; Vancouver
climate change case study 190, **195**,
196–205
public involvement *see* domestic practices;
participation
public spending 127–128, 133

rationalities of experimentation 65–67
renewable energy 178, **180**, 181
research context 1–2; critical gaps 10–11;
distinctiveness of ULL 7–10; emergence
of ULL 5–7; future priorities 254–257;
smart cities 74–75; *see also* design of
ULL; practices of ULL; processes of
ULL; universities
resilience, Manor House case study
107–118
resilient cities 81; *see also* smart cities
risk: 2,000-Watt Society 179, 184–185;
business vs government 193, 199,
204–205; climate change action 122,
204–205; drivers of ULL 24; grassroots
ULL 106; scale of ULL 168
Rotterdam, the Netherlands,
embeddedness case study 235–243

scale: climate change action 193–194, **195**,
198, 202; enabling transformation 120,
122–123; urban living labs 168
Scandinavia, design of ULL 27
second life batteries **180**, 181
security 156, 178
"sense of place" 232–233; *see also*
embeddedness; geographical context
Shenzhen, China 211–220
small and medium enterprises (SMEs)
193
smart cities: China's ULL 220–222, *221*;
comparative analysis 154–160; energy
landscapes 29, 153–154;
experimentation as discursive arena
77–78, 80–82; experimentation as
institutional reconfiguration 77–78,
82–83; experimentation as material site
77–80; practices of ULL 74–76, 83–85,
252; sustainability transitions 76–78;

technological change 147–153, 161;
visions of 148–153, 157–159, 161–162;
see also sustainable cities
smart consumers, energy transitions 58,
64–67
smart grids 53, 54–58, 60–61; *see also*
energy landscapes
social media: citizen science 151–152;
civic steering in China 216; smart cities
147, 148, 159, 161
socio-technical transition: China's ULL
223, *224*; climate change 193, 194–201;
cycling in Greater Manchester 42–49;
governance 3–5; smart cities 75–78
solar power **180**, 181
Stellenbosch: collaborative governance
91–100; innovation and
experimentation 100–103
strategic niche management 3–4, 144–145,
252
strategic ULL **8**, 9, 223, 227
STRIVE programme 128
SubUrbanLab **23**, 29, **31**, 33–34
SusLabNRW 33–34
sustainability transition: 2,000-Watt
Society 176–178; challenges 29; climate
change 191–194, 205; design of ULL
250; domestic practices in Ireland
127–128, 140; embeddedness 232,
233–234, 245; emergence of ULL 5–6;
enabling transformation 118–121;
energy landscapes 52–54, 67–68; smart
cities 76–78
sustainable cities: characteristics of ULL
26–27; design of ULL 22–26, 27–33;
in Europe 21–22, 34–35; governance
22–26; knowledge sharing networks
33–34; *see also* smart cities
Sweden, design of ULL 27
Switzerland *see* 2,000-Watt Society
symbolic understandings of place 233–234

technological change: 2,000-Watt Society
174–176, 179, **180**, 181–183, 185;
energy transitions 52–56, 58–64; smart
cities 147–153, 161; *see also* information
and communication technology
test bed urbanism 61–64
theoretical context of ULL 2, 11
theoretical eclecticism 11, 248

transformation processes: enabling
106–107, 118–123; Manor House
resilience case study 107–118
transition *see* socio-technical transition;
sustainability transition
translation techniques of ULL 234–235,
254
transportation: 2,000-Watt Society
175–176, **180**; cycling in Greater
Manchester 37–38, 39–49; smart cities
82

United Kingdom: cycling in Greater
Manchester 37–38, 39–49; Manor
House resilience case study 107–118
United States, Austin, Texas, energy
transition 54–58, 59–67, 250
universities: 2,000-Watt Society 172, *172*,
183–184; European cities 25; smart grid
in Austin, Texas 54–58
urban areas *see* cities
urban imaginaries 255
urbanism: Manor House resilience case
study 111–118; processes of ULL
106–107
urban living labs (ULL): consequences
of 2, 11; critical gaps 10–11;
distinctiveness 2, 7–10, **25**, 26;
embeddedness 232–235; emergence o
f 1–2, 5–7; experimentation
characteristic 26, 168; meaning of 21–
22, 167–168; multiples modes of 5–10;
scale 168; theoretical context 2, 11;
see also design of ULL; practices of
ULL; processes of ULL
URB@EXP **23**, 29, **31–32**

Vancouver climate change case study:
comparative analysis 201–206;
experimentation in urban spaces
191–194; governance 190–191;
socio-technical climate innovators 193,
194–201, 253–254
Veerkracht Carnisse, Rotterdam, the
Netherlands, embeddedness case study
235–243
vehicles *see* transportation

water governance, HomeLabs project
132–143

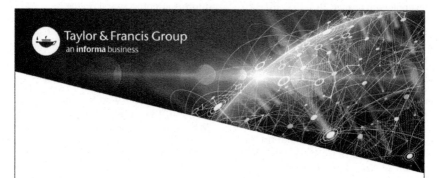